Building a Second Brain 2022

Step by Step Guide to Organize Your Digital Life and Unlock Your Creative Potential

Introduction

The Promise of a Second Brain

How often have you tried to remember something important and felt it slip through your mental grasp?

Perhaps you were having a conversation and couldn't remember a fact that would have convincingly supported your point of view. Maybe you conceived of a brilliant new idea while driving or in transit, but by the time you arrived at your destination, it had evaporated. How often have you struggled to recall even one useful takeaway from a book or article you read in the past?

As the amount of information we have access to grows, such experiences are becoming more and more common. We're flooded with more advice than ever promising to make us smarter, healthier, and happier. We consume more books, podcasts, articles, and videos than we could possibly absorb. What do we really have to show for all the knowledge we've gained? How many of the great ideas we've had or encountered have faded from our minds before we even had a chance to put them into practice?

We spend countless hours reading, listening to, and watching other people's opinions about what we should do, how we should think, and how we should live, but make comparatively little effort applying that knowledge and making it our own. So much of the time we are "information hoarders," stockpiling endless amounts of well-intentioned content that only ends up increasing our anxiety.

This book is dedicated to changing that. You see, all the content you consume online and through all the different kinds of media you have at your disposal isn't useless. It's incredibly important and valuable. The only problem is that you're often consuming it at the wrong time.

What are the chances that the business book you're reading is exactly what you need right at this moment? What are the odds that every single insight from a podcast interview is immediately actionable? How many of the emails sitting in your inbox actually require your full attention right now? More likely, some of it will be relevant now, but most of it will become relevant only at some point in the future.

To be able to make use of information we value, we need a way to package it up and send it through time to our future self. We need a way to cultivate a body of knowledge that is uniquely our own, so when the opportunity arises—whether changing jobs, giving a big presentation, launching a new product, or starting a business or a family—we will have access to the wisdom we need to make good decisions and take the most effective action. It all begins with the simple act of writing things down.

I'll show you how this simple habit is the first step in a system I've developed called Building a Second Brain, which draws on recent advancements in the field of PKM—or personal knowledge management.[1] In the same way that personal computers revolutionized our relationship with technology, personal finance changed how we manage our money, and personal productivity reshaped how we work, personal knowledge management helps us harness the full potential of what we know. While innovations in technology and a new generation of powerful apps have created new opportunities for our times, the lessons you will find within these pages are built on timeless and unchanging principles.

The Building a Second Brain system will teach you how to:

- Find anything you've learned, touched, or thought about in the past within seconds.

- Organize your knowledge and use it to move your projects and goals forward more consistently.

- Save your best thinking so you don't have to do it again.

- Connect ideas and notice patterns across different areas of your life so you know how to live better.

- Adopt a reliable system that helps you share your work more confidently and with more ease.

- Turn work "off" and relax, knowing you have a trusted system keeping track of all the details.

- Spend less time looking for things, and more time doing the best, most creative work you are capable of.

When you transform your relationship to information, you will begin to see the technology in your life not just as a storage medium but as a *tool for thinking*. Like a bicycle for the mind,[II] once we learn how to use it properly, technology can enhance our cognitive abilities and accelerate us toward our goals far faster than we could ever achieve on our own.

In this book I will teach you how to create a system of knowledge management, or a "Second Brain."[III] Whether you call it a "personal cloud," "field notes," or an "external brain" as some of my students have done, it is a digital archive of your most valuable memories, ideas, and knowledge to help you do your job, run your business, and manage your life without having to keep every detail in your head. Like a personal library in your pocket, a Second Brain enables you to recall everything you might want to remember so you can achieve anything you desire.

I've come to believe that personal knowledge management is one of the most fundamental challenges—as well as one of the most incredible opportunities—in the world today. Everyone is in desperate need of a system to manage the ever-increasing volume of information pouring into their brains. I've heard the plea from students and executives, entrepreneurs and managers, engineers and writers, and so many others seeking a more productive and empowered relationship with the information they consume.

Those who learn how to leverage technology and master the flow of information through their lives will be empowered to accomplish anything they set their minds to. At the same time, those who continue to rely on their fragile biological brains will become ever more overwhelmed by the explosive growth in the complexity of our lives.

I've spent years studying how prolific writers, artists, and thinkers of the past managed their creative process. I've spent countless hours researching how human beings can use technology to extend and enhance our natural cognitive abilities. I've personally experimented with every tool, trick, and technique available today for making sense of information. This book distills the very best insights I've discovered from teaching thousands of people around the world how to realize the potential of their ideas.

With a Second Brain at your fingertips, you will be able to unlock the full potential of your hidden strengths and creative instincts. You will have a system that supports you when you are forgetful and unleashes you when you are strong. You will be able to do and learn and create so much more, with so much less effort and stress, than was ever possible before.

In the next chapter, I'll tell you the story of how I built my own Second Brain, and the lessons I learned along the way about how you can build one for yourself.

I. The field of PKM emerged in the 1990s to help university students handle the huge volume of information they suddenly had access to through Internet-connected libraries. It is the individual counterpart to Knowledge Management, which studies how companies and other organizations make use of their knowledge.

II. This metaphor was first used by Steve Jobs to describe the future potential of the personal computer.

III. Other popular terms for such a system include *Zettelkasten* (meaning "slip box" in German, coined by influential sociologist Niklas Luhmann), *Memex* (a word invented by American inventor Vannevar Bush), and *digital garden* (named by popular online creator Anne-Laure Le Cunff).

PART ONE

The Foundation
Understanding What's Possible

Chapter 1

Where It All Started

Your mind is for having ideas, not holding them.
—David Allen, author of *Getting Things Done*

One spring day during my junior year of college, for no apparent reason, I began to feel a small pain in the back of my throat.

I thought it was the first sign of a flu coming on, but my doctor couldn't find a trace of illness. It slowly got worse over the next few months, and I began to visit other, more specialized doctors. They all arrived at the same conclusion: there's nothing wrong with you.

Yet my pain continued getting worse and worse, with no remedy in sight. Eventually it became so severe that I had trouble speaking or swallowing or laughing. I did every diagnostic test and scan imaginable, desperately looking for answers for why I was feeling this way.

As months and then years passed, I began to lose hope that I would ever find relief. I started taking a powerful anti-seizure medication that temporarily relieved the pain, but there were terrible side effects, including a numbing sensation throughout my body and severe short-term memory loss. Entire trips I took, books I read, and precious experiences with loved ones during this period were wiped from my memory as if they never happened. I was a twenty-four-year-old with the mind of an eighty-year-old.

As my ability to express myself continued to deteriorate, my discouragement turned to despair. Without the ability to speak freely, so much of what life had to offer—friendships, dating, traveling, and finding a career I was passionate about—seemed like it was slipping away from me. It felt like a dark curtain was being drawn over the stage of my life before I even had a chance to start my performance.

A Personal Turning Point—Discovering the Power of Writing Things Down

One day, sitting in yet another doctor's office waiting for yet another visit, I had an epiphany. I realized in a flash that I was at a crossroads. I could either take responsibility for my own health and my own treatment from that day forward, or I would spend the rest of my life shuttling back and forth between doctors without ever finding resolution.

I took out my journal and began to write what I was feeling and thinking. I wrote out the history of my condition, through my own lens and in my own words, for the first time. I listed which treatments had helped and which hadn't. I wrote down what I wanted and didn't want, what I was willing to sacrifice and what I wasn't, and what it would mean to me to escape the world of pain I felt trapped within.

As the story of my health began to take shape on the page, I knew what I needed to do. I stood up abruptly, walked over to the receptionist, and asked for my complete patient record. She looked at me quizzically, but after I answered a few questions, she turned to her files and began making photocopies.

My patient record amounted to hundreds of pages, and I knew I would never be able to keep track of them on paper. I started scanning every page on my family's home computer, turning them into digital records that could be searched, rearranged, annotated, and shared. I became the project manager of my own condition, taking detailed notes on everything my doctors told me, trying out every suggestion they made, and generating questions to review during my next appointment.

With all this information in one place, patterns began to emerge. With my doctors' help, I discovered a class of afflictions called "functional voice disorders," which included problems with any of the more than fifty pairs of muscles required to properly swallow a piece of food. I realized that the medications I was taking were masking my symptoms, and in the process making it harder to hear what they were telling me. What I had was not an illness or infection that could be eradicated with a pill—it was a functional condition that required changes in how I took care of my body.

I began to research how breathing, nutrition, vocal habits, and even past experiences in childhood can be manifested in the nervous system. I started to understand the mind-body connection and how my thoughts and feelings directly impacted the way my body felt. Taking notes on everything I learned, I devised an experiment: I would try a few simple lifestyle changes, such as improving my diet and regular meditation, combined with a series of voice exercises I learned from a voice therapist. To my shock and amazement, it began to work almost immediately. My pain didn't disappear, but it became far more manageable.[1]

As I look back, my notes were as important in finding relief as any medicine or procedure. They gave me the chance to step back from the details of my condition and see my situation from a different perspective. For both the outer world of medicine and the inner world of sensations, my notes were a practical medium for turning any new information I encountered into practical solutions I could use.

From then on, I became obsessed with the potential of technology to channel the information all around me. I began to realize that the simple act of taking notes on a computer was the tip of an iceberg. Because once made digital, notes were no longer limited to short, handwritten scribbles—they could take any form, including images, links, and files of any shape and size. In the digital realm, information could be molded and shaped and directed to any purpose, like a magical, primordial force of nature.

I started using digital notetaking in other parts of my life. In my college classes, I turned stacks of disheveled spiral-bound notebooks into an elegant, searchable collection of lessons. I learned to master the process of writing down only the most important points from my classes, reviewing them on demand, and using them to compose an essay or pass a test. I had always been a mediocre student with average grades. My early schoolteachers would regularly send me home with report cards noting my short attention span and wandering mind. You can imagine my delight when I graduated from college with a nearly straight-A grade point average and university honors.

I had the misfortune of graduating into one of the worst job markets in a generation, in the aftermath of the 2008 financial crisis. Faced with few employment opportunities in the United States, I decided to join the Peace Corps, an overseas volunteer program that sends Americans to serve in developing countries. I was accepted and assigned to a small school in the eastern Ukrainian countryside, where I would spend two years teaching English to students aged eight to eighteen.

Working as a teacher with few resources and little support, my notetaking system once again became my lifeline. I saved examples of lessons and exercises anywhere I found them: from textbooks, websites, and USB drives passed around by other teachers. I mixed and matched English phrases, expressions, and slang into word games to keep my energetic third graders engaged. I taught the older students the basics of personal productivity—how to keep a schedule, how to take notes in class, and how to set goals and plan their education. I will never forget their appreciation as they grew up and used those skills to apply to universities and succeed in their first jobs. Years later, I still regularly receive messages of gratitude as the productivity skills I taught my former students continue to bear fruit in their lives.

I returned to the US after two years of service and was thrilled to land a job as an analyst at a small consulting firm in San Francisco. As excited as I was to start my career, I was also faced with a major challenge: the pace of work was frantic and overwhelming. Moving straight from rural Ukraine to the epicenter of Silicon Valley, I was utterly unprepared for the constant barrage of inputs that is a normal part of modern workplaces. Every day I received hundreds of emails, every hour dozens of messages, and the pings and dings from every device merged into a ceaseless melody of interruption. I remember looking around at my colleagues and wondering, "How can anyone get anything done here? What's their secret?"

I knew only one trick, and it started with writing things down.

I started taking notes on everything I was learning using a notetaking app on my computer. I took notes during meetings, on phone calls, and while doing research online. I wrote down facts gleaned from research papers that could be used in the slides we presented to clients. I wrote down tidbits of insight I came across on social media, to share on our own social channels. I wrote down feedback from my more experienced colleagues so I could make sure I digested it and took it to heart. Every time we started a new project, I created a dedicated place on my computer for the information related to it, where I could sort through it all and decide on a plan of action.

As the information tide receded, I started to gain a sense of confidence in my ability to find exactly what I needed when I needed it. I became the go-to person in the office for finding that one file, or unearthing that one fact, or remembering exactly what the client had said three weeks earlier. You know the feeling of satisfaction when you are the only one in the room who remembers an important detail? That feeling became the prize in my personal pursuit to capitalize on the value of what I knew.

Another Shift—Discovering the Power of Sharing

My collection of notes and files had always been for my own personal use, but as I worked on consulting projects for some of the most important organizations in the world, I started to realize that it could be a business asset as well.

I learned from one of the reports we published that the value of physical capital in the US—land, machinery, and buildings for example—is about $10 trillion, but that value is dwarfed by the total value of *human* capital, which is estimated to be five to ten times larger. Human capital includes "the knowledge and the knowhow embodied in humans—their education, their experience, their wisdom, their skills, their relationships, their common sense, their intuition."[1]

If that was true, was it possible that my personal collection of notes was a knowledge asset that could grow and compound over time? I began to see my as-yet-unnamed Second Brain not just as a notetaking tool but as a loyal confidant and thought partner. When I was forgetful, it always remembered. When I lost my way, it reminded me where we were going. When I felt stuck and at a loss for ideas, it suggested possibilities and pathways.

At one point some of my colleagues asked me to teach them my organizing methods. I found that virtually all of them already used various productivity tools, such as paper notepads or the apps on their smartphones, but that very few did so in a systematic, intentional way. They tended to move information around from place to place haphazardly, reacting to the demands of the moment, never quite trusting that they'd be able to find it again. Every new productivity app promised a breakthrough, but usually ended up becoming yet another thing to manage.

Casual lunchtime chats with my colleagues turned into a book club, which became a workshop, which eventually evolved into a paid class open to the public. As I taught what I knew to more and more people and saw the immediate difference it made in their work and lives, it began to dawn on me that I had discovered something very special. My experience managing my chronic condition had taught me a way of getting organized that was ideal for solving problems and producing results now, not in some far-off future. Applying that approach to other areas of my life, I had found a way to organize information holistically—for a variety of purposes, for any project or goal—instead of only for one-off tasks. And more than that, I discovered that once I had that information at hand, I could easily and generously share it in all kinds of ways to serve the people around me.

The Origins of the Second Brain System

I began to call the system I had developed my Second Brain and started a blog to share my ideas about how it worked. These ideas resonated with a much wider audience than I ever expected, and my work was eventually featured in publications such as the *Harvard Business Review*, *The Atlantic*, *Fast Company*, and *Inc.*, among others. An article I wrote about how to use digital notetaking to enhance creativity went viral in the productivity community, and I was invited to speak and teach workshops at influential companies like Genentech, Toyota, and the Inter-American Development Bank. In early 2017, I decided to create an online course called "Building a Second Brain" to teach my system on a wider scale.[II] In the years since, that program has produced thousands of graduates from more than one hundred countries and every walk of life, creating an engaged and inquisitive community where the lessons in this book have been honed and refined.

In the next couple of chapters, I'll show you how the practice of creating a Second Brain is part of a long legacy of thinkers and creators who came before us—writers, scientists, philosophers, leaders, and everyday people who strived to remember and achieve more. Then I'll introduce you to a few basic principles and tools you'll need to set yourself up to succeed. Part Two, "The Method," introduces each of the four steps you'll follow to build a Second Brain so you can immediately begin to capture and share ideas with more intention. And Part Three, "Making Things Happen," offers a set of powerful ways to use your Second Brain to enhance your productivity, accomplish your goals, and thrive in your work and life.

I've shared my story with you because I want you to know that this book isn't about perfectly optimizing some kind of idealized life. Everyone experiences pain, makes mistakes, and struggles at some point in their lives. I've had my fair share of challenges, but at each stage of my journey, treating my thoughts as treasures worth keeping has been the pivotal element in everything I've overcome and achieved.

You may find this book in the "self-improvement" category, but in a deeper sense it is the opposite of self-improvement. It is about optimizing a *system outside yourself*, a system not subject to your limitations and constraints, leaving you happily unoptimized and free to roam, to wonder, to *wander* toward whatever makes you feel alive here and now in each moment.

I. I was aided in this effort by my involvement with the Quantified Self community, a network of local meetup groups in which people share their stories about how they track their health, productivity, mood, or behavior to learn more about themselves.

II. Interested readers can find out more at buildingasecondbrain.com/course.

Chapter 2

What Is a Second Brain?

We extend beyond our limits, not by revving our brains like a machine or bulking them up like a muscle—but by strewing our world with rich materials, and by weaving them into our thoughts.
—Annie Murphy Paul, author of *The Extended Mind*

Information is the fundamental building block of everything you do.

Anything you might want to accomplish—executing a project at work, getting a new job, learning a new skill, starting a business—requires finding and putting to use the right information. Your professional success and quality of life depend directly on your ability to manage information effectively.

According to the *New York Times*, the average person's daily consumption of information now adds up to a remarkable 34 gigabytes.[1] A separate study cited by the *Times* estimates that we consume the equivalent of 174 full newspapers' worth of content each and every day, five times higher than in 1986.[2]

Instead of empowering us, this deluge of information often overwhelms us. Information Overload has become Information Exhaustion, taxing our mental resources and leaving us constantly anxious that we're forgetting something. Instantaneous access to the world's knowledge through the Internet was supposed to educate and inform us, but instead it has created a society-wide poverty of attention.[1]

Research from Microsoft shows that the average US employee spends 76 hours per year looking for misplaced notes, items, or files.[3] And a report from the International Data Corporation found that 26 percent of a typical knowledge worker's day is spent looking for and consolidating information spread across a variety of systems.[4] Incredibly, only 56 percent of the time are they able to find the information required to do their jobs.

In other words, we go to work five days per week, but spend more than one of those days on average just looking for the information we need to do our work. Half the time, we don't even succeed in doing that.

It's time for us to upgrade our Paleolithic memory. It's time to acknowledge that we can't "use our head" to store everything we need to know and to outsource the job of remembering to intelligent machines. We have to recognize that the cognitive demands of modern life increase every year, but we're still using the same brains as two hundred thousand years ago, when modern humans first emerged on the plains of East Africa.

Every bit of energy we spend straining to recall things is energy not spent doing the thinking that only humans can do: inventing new things, crafting stories, recognizing patterns, following our intuition, collaborating with others, investigating new subjects, making plans, testing theories. Every minute we spend trying to mentally juggle all the stuff we have to do leaves less time for more meaningful pursuits like cooking, self-care, hobbies, resting, and spending time with loved ones.

However, there's a catch: every change in how we use technology also requires a change in how we think. To properly take advantage of the power of a Second Brain, we need a new relationship to information, to technology, and even to ourselves.

The Legacy of Commonplace Books

For insight into our own time, we can look to history for lessons on what worked in other eras. The practice of writing down one's thoughts and notes to help make sense of the world has a long legacy. For centuries, artists and intellectuals from Leonardo da Vinci to Virginia Woolf, from John Locke to Octavia Butler, have recorded the ideas they found most interesting in a book they carried around with them, known as a "commonplace book."[II]

Popularized in a previous period of information overload, the Industrial Revolution of the eighteenth and early nineteenth centuries, the commonplace book was more than a diary or journal of personal reflections. It was a learning tool that the educated class used to understand a rapidly changing world and their place in it.

In *The Case for Books*,[5] historian and former director of the Harvard University Library Robert Darnton explains the role of commonplace books:

> *Unlike modern readers, who follow the flow of a narrative from beginning to end, early modern Englishmen read in fits and starts and jumped from book to book. They broke texts into fragments and assembled them into new patterns by transcribing them in different sections of their notebooks. Then they reread the copies and rearranged the patterns while adding more excerpts. Reading and writing were therefore inseparable activities. They belonged to a continuous effort to make sense of things, for the world was full of signs: you could read your way through it; and by keeping an account of your readings, you made a book of your own, one stamped with your personality.*[III]

Commonplace books were a portal through which educated people interacted with the world. They drew on their notebooks in conversation and used them to connect bits of knowledge from different sources and to inspire their own thinking.

As a society, all of us could benefit from the modern equivalent of a commonplace book. The media landscape of today is oriented toward what is *novel* and *public*—the latest political controversy, the new celebrity scandal, or the viral meme of the day. Resurrecting the commonplace book allows us to stem the tide, shifting our relationship with information toward the *timeless* and the *private*.

Instead of consuming ever-greater amounts of content, we could take on a more patient, thoughtful approach that favors rereading, reformulating, and working through the implications of ideas over time. Not only could this lead to more civil discussions about the important topics of the day; it could also preserve our mental health and heal our splintered attention.

But this isn't simply a return to the past. We now have the opportunity to supercharge the custom of commonplace books for the modern era. We have the chance to turn that historical practice into something far more flexible and convenient.

The *Digital* Commonplace Book

Once our notes and observations become digital, they can be searched, organized and synced across all our devices, and backed up to the cloud for safekeeping. Instead of randomly scribbling down notes on pieces of paper, hoping we'll be able to find them later, we can cultivate our very own "knowledge vault" so we always know exactly where to look.

Writer and photographer Craig Mod wrote, "There is a gaping opportunity to consolidate our myriad marginalia[IV] into an even more robust commonplace book. One searchable, always accessible, easily shared and embedded amongst the digital text we consume."[6]

This digital commonplace book is what I call a Second Brain. Think of it as the combination of a study notebook, a personal journal, and a sketchbook for new ideas. It is a multipurpose tool that can adapt to your changing needs over time. In school or courses you take, it can be used to take notes for studying. At work, it can help you organize your projects. At home, it can help you manage your household.

However you decide to use it, your Second Brain is a private knowledge collection designed to serve a lifetime of learning and growth, not just a single use case. It is a laboratory where you can develop and refine your thinking in solitude before sharing it with others. A studio where you can experiment with ideas until they are ready to be put to use in the outside world. A whiteboard where you can sketch out your ideas and collaborate on them with others.

As soon as you understand that we naturally use digital tools to extend our thinking beyond the bounds of our skulls, you'll start to see Second Brains everywhere.

A calendar app is an extension of your brain's ability to remember events, ensuring you never forget an appointment. Your smartphone is an extension of your ability to communicate, allowing your voice to reach across oceans and continents. Cloud storage is an extension of your brain's memory, allowing you to store thousands of gigabytes and access them from anywhere.[v]

It's time to add digital notes to our repertoire and further enhance our natural capabilities using technology.

Rethinking Notetaking: Notes as Knowledge Building Blocks

In past centuries, only the intellectual elite needed commonplace books—writers, politicians, philosophers, and scientists who had a reason to synthesize their writing or research.

Nowadays, almost everyone needs a way to manage information.

More than half the workforce today can be considered "knowledge workers"—professionals for whom knowledge is their most valuable asset, and who spend a majority of their time managing large amounts of information. In addition, no matter what our formal role is, all of us have to come up with new ideas, solve novel problems, and communicate with others effectively. We have to do these things regularly, reliably, not just once in a while.

As a knowledge worker, where does your knowledge live? Where does your knowledge go when it's created or discovered? "Knowledge" can seem like a lofty concept reserved exclusively for scholars and academics, but at the most practical level, knowledge begins with the simple, time-honored practice of taking notes.

For many people, their understanding of notetaking was formed in school. You were probably first told to write something down because it would be on the test. This implied that the minute the test was over, you would never reference those notes again. Learning was treated as essentially disposable, with no intention of that knowledge being useful for the long term.

When you enter the professional world, the demands on your notetaking change completely. The entire approach to notetaking you learned in school is not only obsolete, it's the exact opposite of what you need.

In the professional world:

- It's not at all clear what you should be taking notes on.
- No one tells you when or how your notes will be used.
- The "test" can come at any time and in any form.
- You're allowed to reference your notes at any time, provided you took them in the first place.
- You are expected to take action on your notes, not just regurgitate them.

This isn't the same notetaking you learned in school. It's time to elevate the status of notes from test prep and humble scribblings into something far more interesting and dynamic. For modern, professional notetaking, a note is a "knowledge building block"—a discrete unit of information interpreted through your unique perspective and stored outside your head.

By this definition, a note could include a passage from a book or article that you were inspired by; a photo or image from the web with your annotations; or a bullet-point list of your meandering thoughts on a topic, among many other examples. A note could include a single quote from a film that really struck you, all the way to thousands of words you saved from an in-depth book. The length and format don't matter—if a piece of content has been interpreted through your lens, curated according to your taste, translated into your own words, or drawn from your life experience, and stored in a secure place, then it qualifies as a note.

A knowledge building block is discrete. It stands on its own and has intrinsic value, but knowledge building blocks can also be combined into something much greater—a report, an argument, a proposal, a story.

Like the LEGO blocks you may have played with as a kid, they can be rapidly searched, retrieved, moved around, assembled, and reassembled into new forms without requiring you to invent anything from scratch. You need to put in the effort to create a note only once, and then you can just mix and match and try out different combinations until something clicks.

Technology doesn't just make notetaking more efficient. It transforms the very nature of notes. No longer do we have to write our thoughts on Post-its or notepads that are fragile, easy to lose, and impossible to search. Now we write notes in the cloud, and the cloud follows us everywhere. No longer do we have to spend countless hours meticulously cataloging and transcribing our thoughts on paper. Now we collect knowledge building blocks and spend our time imagining the possibilities for what they could become.

A Tale of Two Brains

Let me paint a picture of a day in the life of someone who doesn't have a Second Brain, and someone who does. See if either of these descriptions sounds familiar.

Nina wakes up on Monday morning, and before her eyes even open, thoughts are flooding her brain. Things to do, things to think about, things to decide. It all comes rushing in from the depths of her subconscious, where it's been simmering all weekend.

Nina's thoughts continue to swirl around her brain as she gets ready for work. Like jittery birds, they flit and flutter around her head because they have nowhere else to rest. There is a constant hum of background anxiety that she has come to expect, as she wonders what needs her attention and what she may be missing.

After a hectic morning, Nina finally sits down at her desk to start her workday, opens up her email inbox, and is instantly engulfed by a torrent of new messages. Flashing with urgent subject lines and the names of important senders, these demands fill her with a cold adrenaline rush. She knows that her morning is shot, her own plans ruined. Pushing aside the important work she wanted to focus on this morning, Nina settles in for a long slog of replying to emails.

By the time she gets back from lunch, Nina is finally done handling the most urgent issues. It's finally time to focus on the priorities she's set for herself. This is when the reality sets in: after a morning spent fighting fires, she's far too scatterbrained and tired to focus. Like so many times before, Nina lowers her expectations, settling for chipping away slowly at her ever-expanding to-do list full of other people's priorities.

After work, Nina has one last chance to work on the project that she knows will make use of her talents and take her career to the next level. She exercises, eats dinner, and spends some quality time with the kids. As they go to bed, she's filled with enthusiasm that she finally gets some time to herself.

She sits down at the computer, and the questions begin: "Where did I leave off last time? Where did I put that file? Where are all my notes?"

By the time Nina gets set up and ready to go, she's far too tired to make real progress. This pattern repeats itself day after day. After enough of these false starts, she starts to give up. Why even try? Why keep attempting to do the impossible? Why resist the temptation to watch another Netflix episode or scroll through social media? Without the time and energy to move things decisively forward, what's the point of starting?

Nina is a competent, responsible, and hardworking professional. Many people would feel privileged to be in her shoes. There's nothing wrong with the work she does or the life she leads, yet underneath the respectable exterior, there is something missing. She isn't meeting her own standards for what she knows she's capable of. There are experiences that she wants for herself and her family that seem to continuously get postponed, waiting for "someday" when somehow she will have the time and space to make them happen.

Does anything about Nina's experience sound familiar? Every detail of her story is real, drawn from messages people have sent me over the years. Their stories convey a pervasive feeling of discontent and dissatisfaction—the experience of facing an endless onslaught of demands on their time, their innate curiosity and imagination withering away under the suffocating weight of obligation.

So many of us share the feeling that we are surrounded by knowledge, yet starving for wisdom. That despite all the mind-expanding ideas we have access to, the quality of our attention is only getting worse. That we are paralyzed by the conflict between our responsibilities and our most heartfelt passions, so that we're never quite able to focus and also never quite able to rest.

There is an alternative story. A different way a Monday morning can go. It is also drawn from the real-life stories I've received, this time from people who have built a Second Brain for themselves.

You wake up Monday morning, looking forward to starting your day and your week. As you get out of bed, take a shower, and get dressed, the thoughts start arriving. You have just as many worries and responsibilities as anyone else, but you also have a secret weapon.

In the shower, you suddenly realize there's a better way to advance the project you're focused on at work. As you step out onto the mat, you jot down the idea as a digital note on your smartphone. Over breakfast with your family, you find your mind already working out the new strategy, pondering its implications. Those thoughts get jotted down as well, in the brief moments between feeding the kids and sending them off to school. As you drive to work, you start realizing there are challenges you haven't considered. You dictate a quick audio memo to your phone as you drive, which gets automatically transcribed and saved in your notes.

Monday morning in the office is the usual whirlwind, with emails and chat messages and phone calls arriving at their usual frantic pace. As you share your new idea with your colleagues, they start asking questions, pointing out valid concerns, and adding their own contributions. At each of these moments, you are ready to save them as notes in your Second Brain. You withhold judgment, seeking to gather the widest possible range of feedback before deciding on a course of action.

Before you know it, it's lunchtime. As you take a break to grab a bite to eat, your thoughts turn philosophical: "What is the ultimate point of the project, and are we forgetting it? How does it fit into the long-term vision of the product we want to build? What is the impact of the new strategy on shareholders, customers, suppliers, and the environment?" You have only thirty minutes to eat lunch, and you don't have time to ponder these questions in depth, but you note them down as a reminder to think about later.

You are on your smartphone just like everyone else, but you aren't doing what they are doing. You are creating value instead of killing time.

By the time the afternoon meeting comes around to review the strategy you've come up with, you already have a formidable collection of notes ready and waiting: the ideas, strategies, objectives, challenges, questions, concerns, contributions, and reminders you've collected over just a few hours on a Monday morning.

You take ten minutes before the meeting starts to organize your notes. About a third of them aren't a priority, and you put them aside. Another third are critical, and you make them into an agenda for the meeting. The remaining third are somewhere in between, and you put them into a separate list to refer to if appropriate.

As the meeting begins, the team sits down to start discussing the project. You are already prepared. You've already considered the biggest problems from several different angles, mapped out a number of possible solutions, and started thinking about the big-picture implications. You've even received feedback from some of your colleagues and incorporated it into your recommendations. You argue for your point of view while also remaining open to the perspectives of your team. Your goal is to stay present and guide the conversation to the best possible outcome, making use of everyone's unique way of seeing things. All the important reflections, new ideas, and unexpected possibilities your colleagues come up with also get recorded in your Second Brain.

As this way of working with information continues over days and weeks and months, the way your mind works begins to change. You start to see recurring patterns in your thinking: why you do things, what you really want, and what's really important to you. Your Second Brain becomes like a mirror, teaching you about yourself and reflecting back to you the ideas worth keeping and acting on. Your mind starts to become intertwined with this system, leaning on it to remember more than you ever could on your own.

All this is literally not just in your head. People can tell there is something different about you. They start to recognize that you can draw on an unusually large body of knowledge at a moment's notice. They remark on your amazing memory, but what they don't know is that you never even try to remember anything. They admire your incredible dedication to developing your thinking over time. In reality, you are just planting seeds of inspiration and harvesting them as they flower.

As you begin to see all the knowledge you've gained in tangible form, it dawns on you that you already have everything you need to strike out toward the future you want. There's no need to wait until you're perfectly prepared. No need to consume more information or do more research. All that's left is for you to take action on what you already know and already have, which is laid out before you in meticulous detail.

Your brain is no longer the bottleneck on your potential, which means you have all the bandwidth you need to pursue any endeavor and make it successful. This sense of confidence in the quality of your thinking gives you the freedom to ask deeper questions and the courage to pursue bigger challenges. You can't fail, because failure is just more information, to be captured and used as fuel for your journey.

This is what it's like to build and harness the power of a Second Brain.

Leveraging Technology as Thinking Tools

Throughout the twentieth century, a series of scholars and innovators[7] offered a vision for how technology could change humanity for the better. They dreamed of creating an "extended mind" that would amplify human intellect and help us solve the greatest problems facing society.[VI] The possibility of such a technological marvel shined like a beacon for the future, promising to liberate knowledge from dusty old books and make it universally accessible and useful.[VII]

Their efforts were not in vain. Those ideas inspired much of the technology that we use every day, but paradoxically, despite all the technological inventions of the Information Age, we are in some ways further from their original vision than ever. We spend hours every day interacting with social media updates that will be forgotten in minutes. We bookmark articles to read later, but rarely find the time to revisit them again. We create documents that are used once and then get abandoned in the abyss of our email or file systems. So much of our intellectual output—from brainstorms to photos to planning to research—all too often is left stranded on hard drives or lost somewhere in the cloud.

I believe that we have reached an inflection point, where technology has become sufficiently advanced and user-friendly that we can integrate it with our biological brains. Computers have become smaller, more powerful, and more intuitive, to the point that they are unmistakably an essential component of how we think.

The time has come for us to realize the vision of technology's early pioneers—that everyone should have an extended mind not just to remember more and be more productive, but to lead more fulfilling lives.

I. Herbert Simon, an American economist and cognitive psychologist, wrote, "What information consumes is rather obvious: it consumes the attention of its recipients. Hence a wealth of information creates a poverty of attention…"

II. The word "commonplace" can be traced back to Ancient Greece, where a speaker in law courts or political meetings would keep an assortment of arguments in a "common place" for easy reference.

III. The practice of keeping personal notes also arose in other countries, such as *biji* in China (roughly translated as "notebook"), which could contain anecdotes, quotations, random musings, literary criticism, short fictional stories, and anything else that a person thought worth recording. In Japan, *zuihitsu* (known as "pillow books") were collections of notebooks used to document a person's life.

IV. "Marginalia" refers to the marks made in the margins of a book or other document, including scribbles, comments, annotations, critiques, doodles, or illustrations.

V. Have you ever lost your smartphone or been unable to access the Internet, and felt like a critical part of yourself was missing? That's a sign that an external tool has become an extension of your mind. In a 2004 study, Angelo Maravita and Atsushi Iriki discovered that when monkeys and humans consistently use a tool to extend their reach, such as using a rake to reach an object, certain neural networks in the brain change their "map" of the body to include the new tool. This fascinating finding reinforces the idea that external tools can and often do become a natural extension of our minds.

VI. Recent advancements and discoveries in the field of "extended cognition" have shed new light on how practical and powerful it can be to "think outside the brain." This book isn't focused on the science, but for an excellent introduction to extended cognition I recommend *The Extended Mind* by Annie Murphy Paul.

VII. Vannevar Bush wrote of a "scholar's workstation" called a "Memex," which was "a device in which an individual stores all his books, records, and communications, and which is mechanized so that it may be consulted with exceeding speed and flexibility. It is an enlarged intimate supplement to his memory."

Chapter 3

How a Second Brain Works

It is in the power of remembering that the self's ultimate freedom consists. I am free because I remember.
—Abhinavagupta, tenth-century Kashmiri philosopher and mystic

Think of your Second Brain as the world's best personal assistant.

It is perfectly reliable and totally consistent. It is always ready and waiting to capture any bit of information that might be of value to you. It follows directions, makes helpful suggestions, and reminds you of what's important to you.

What would the job description for such a personal assistant look like? What "jobs" would you hire them to do for you? The same way you would hold your assistant accountable to a certain standard of performance, the same is true for your Second Brain. You need to know what it should be doing for you so you know if it's worth keeping around.

In this chapter we'll see how the four main capabilities of a Second Brain will actively work for you—immediately and over time; the one basic tool you'll need to get started; how your Second Brain will evolve to serve what is most important to you; and, finally, an introduction to the four steps of the CODE Method that lies at the heart of it all.

The Superpowers of a Second Brain

There are four essential capabilities that we can rely on a Second Brain to perform for us:

1. Making our ideas concrete.

2. Revealing new associations between ideas.

3. Incubating our ideas over time.

4. Sharpening our unique perspectives.

Let's examine each of these.

Second Brain Superpower #1: Make Our Ideas Concrete

Before we do anything with our ideas, we have to "off-load" them from our minds and put them into concrete form. Only when we declutter our brain of complex ideas can we think clearly and start to work with those ideas effectively.

In 1953, American biologist James Watson and English physicist Francis Crick made a profound discovery: the structure of DNA was a double helix. Their discovery was built on the groundwork laid by other pioneers, including advancements in X-ray crystallography by Rosalind Franklin and Maurice Wilkins, and ushered in a golden age in molecular biology and genetics.

Watson and Crick's breakthrough is well recognized, but there is a part of their story that is much less well known. An important tool of the researchers was building physical models, an approach they borrowed from American biochemist Linus Pauling. They made cardboard cutouts to approximate the shapes of the molecules they knew were part of DNA's makeup and, like a puzzle, experimented with different ways of putting them together. They would shift around their models on their desktops, trying to find a shape that fit everything they knew about how the molecules were arranged. The double helix structure seemed to fit all known constraints, allowing the complementary base pairs to fit together perfectly while respecting the ratios between elements that had been measured previously.[1]

This is a remarkable aspect of one of the most famous scientific discoveries of the last century: at the decisive moment, even highly trained scientists deeply familiar with mathematical and abstract thinking turned to the most basic, ancient medium available: physical stuff.

Digital notes aren't physical, but they are visual. They turn vague concepts into tangible entities that can be observed, rearranged, edited, and combined together. They may exist only in virtual form, but we can still see them with our eyes and move them around with our fingers. As researchers Deborah Chambers and Daniel Reisberg found in their research on the limits of mental visualization, "The skills we have developed for dealing with the external world go beyond those we have for dealing with the internal world."[2]

Second Brain Superpower #2: Reveal New Associations Between Ideas

In its most practical form, creativity is about connecting ideas together, especially ideas that don't seem to be connected.

Neuroscientist Nancy C. Andreasen, in her extensive research on highly creative people including accomplished scientists, mathematicians, artists, and writers, came to the conclusion that "Creative people are better at recognizing relationships, making associations and connections."[3]

By keeping diverse kinds of material in one place, we facilitate this connectivity and increase the likelihood that we'll notice an unusual association.

Quotes from a philosophy book written in ancient times might sit next to the latest clever tweet. Screenshots from an interesting YouTube video can live right by scenes from classic movies. An audio memo might be saved alongside project plans, a link to a helpful website, and a PDF with the latest research findings. All these formats can be combined in a way that would be impossible in the physical world.

If you've ever played the word-tile game Scrabble, you know the best way to come up with new words is to mix up the letters in different combinations until a word jumps out at you. In our Second Brain we can do the same: mix up the order of our ideas until something unexpected emerges. The more diverse and unusual the material you put into it in the first place, the more original the connections that will emerge.

Second Brain Superpower #3: Incubate Our Ideas Over Time

Too often when we take on a task—planning an event, designing a product, or leading an initiative—we draw only on the ideas we have access to right in that moment. I call this approach a "heavy lift"—demanding instantaneous results from our brains without the benefit of a support system.

Even when we do a brainstorm, that still relies only on ideas that we can think up right now. What are the chances that the most creative, most innovative approaches will instantly be top of mind? What are the odds that the best way to move forward is one of the first ways we come up with?

This tendency is known as *recency bias*.[4] We tend to favor the ideas, solutions, and influences that occurred to us most recently, regardless of whether they are the best ones. Now imagine if you were able to unshackle yourself from the limits of the present moment, and draw on weeks, months, or even years of accumulated imagination.

I call this approach the "slow burn"—allowing bits of thought matter to slowly simmer like a delicious pot of stew brewing on the stove. It is a calmer, more sustainable approach to creativity that relies on the gradual accumulation of ideas, instead of all-out binges of manic hustle. Having a Second Brain where lots of ideas can be permanently saved for the long term turns the passage of time into your friend, instead of your enemy.

Second Brain Superpower #4: Sharpen Our Unique Perspectives

Until now we've talked mostly about gathering the ideas of others, but the ultimate purpose of a Second Brain is to allow your own thinking to shine.

A recent study from Princeton University found that there is a certain kind of job that is least likely to be automated by machines in coming years. Surprisingly, it wasn't jobs that required advanced skills or years of training that were predicted to fare best. It was jobs that required the ability to convey "not just information but a particular interpretation of information."[5]

In other words, the jobs that are most likely to stick around are those that involve promoting or defending a particular perspective. Think of a fundraising organizer sharing stories of the impact their nonprofit has made, a researcher using data to back up their interpretation of an experiment, or a project manager citing a couple of key precedents to support a decision. Our careers and businesses depend more than ever on our ability to advance a particular point of view and persuade others to adopt it as well.[6]

Advocating for a particular point of view isn't just a matter of sparkling charisma or irresistible charm. It takes supporting material.

American journalist, author, and filmmaker Sebastian Junger once wrote on the subject of "writer's block": "It's not that I'm blocked. It's that I don't have enough research to write with power and knowledge about that topic. It always means, not that I can't find the right words, [but rather] that I don't have the ammunition."[7]

When you feel stuck in your creative pursuits, it doesn't mean that there's something wrong with you. You haven't lost your touch or run out of creative juice. It just means you don't yet have enough raw material to work with. If it feels like the well of inspiration has run dry, it's because you need a deeper well full of examples, illustrations, stories, statistics, diagrams, analogies, metaphors, photos, mindmaps, conversation notes, quotes—anything that will help you argue for your perspective or fight for a cause you believe in.

Choosing a Notetaking App: The Neural Center of Your Second Brain

The same technology that has fueled an explosion in the volume of information coming our way has also provided the tools to help us manage it.

While your Second Brain is made up of all the tools you use to interact with information, including a to-do list, a calendar, email, and reading apps, for example, there is one category of software I recommend as the centerpiece of your Second Brain: a digital notetaking app.[1] There are plentiful options available, from the free notes app that came preinstalled on your smartphone to third-party software you can download with exactly the features you need.

From Microsoft OneNote, Google Keep, and Apple Notes to Notion and Evernote, digital notes apps have four powerful characteristics that make them ideal for building a Second Brain. They are:

- **Multimedia:** Just like a paper notebook might contain drawings and sketches, quotes and ideas, and even a pasted photo or Post-it, a notes app can store a wide variety of different kinds of content in one place, so you never need to wonder where to put something.

- **Informal:** Notes are inherently messy, so there's no need for perfect spelling or polished presentation. This makes it as easy and frictionless as possible to jot things down as soon as they occur to you, which is essential to allow nascent ideas to grow.

- **Open-ended:** Taking notes is a continuous process that never really ends, and you don't always know where it might lead. Unlike more specialized kinds of software that are designed to produce a specific kind of output (such as slide decks, spreadsheets, graphics, or videos), notes are ideal for free-form exploration before you have a goal in mind.

- **Action-oriented:** Unlike a library or research database, personal notes don't need to be comprehensive or precise. They are designed to help you quickly capture stray thoughts so you can remain focused on the task at hand.

All four of the above qualities are shared by paper notes, but when we make them digital, we can supercharge these timeless benefits with the incredible capabilities of technology—searching, sharing, backups, editing, linking, syncing between devices, and many others. Digital notes combine the casual artistry of a daily sketchbook with the scientific power of modern software.

Choosing which apps and tools you'll use is a personal decision, and depends on which mobile device you use, the needs of your job or business, and even your own temperament and taste. The software landscape is dynamic and constantly changing. New apps regularly come out, and existing ones release innovative new features. You can find a free, continually updated guide to choosing your notes app and other Second Brain tools at Buildingasecondbrain.com/resources.

Although you will always use many different software programs to manage information and do your work—from word processors to messaging platforms and project management tools used within your organization—your notes app is uniquely designed to facilitate personal knowledge management.

A good place to start is to look at the apps you already have and perhaps are already using. You can always start now with a basic option and upgrade later as your needs get more sophisticated.[II]

Most important of all, don't get caught in the trap of perfectionism: insisting that you have to have the "perfect" app with a precise set of features before you take a single note. It's not about having the perfect tools—it's about having a *reliable* set of tools you can depend on, knowing you can always change them later.

Remembering, Connecting, Creating: The Three Stages of Personal Knowledge Management

As people set out on their Second Brain journey, there are three stages of progress I often observe—and even encourage. Those stages are *remembering*, *connecting*, and *creating*. It takes time to fully unlock the value of using digital tools to enhance and extend what our minds are capable of, but there are also distinct benefits at every step along the way.

The first way that people tend to use their Second Brain is as a memory aid. They use their digital notes to save facts and ideas that they would have trouble recalling otherwise: takeaways from meetings, quotes from interviews, or the details of a project, for example.

Camille is the cofounder and lead designer of a start-up in Quebec, Canada. She uses her Second Brain to save excerpts from the many research reports and studies she reads as part of her work designing electric vehicle charging stations for large residential buildings. Most of these reports are published as PDFs, a notoriously inflexible and difficult-to-use format, but by importing the findings most relevant to her work into her notes, she can add as many annotations and comments as she wants.

The second way that people use their Second Brain is to connect ideas together. Their Second Brain evolves from being primarily a *memory* tool to becoming a *thinking* tool. A piece of advice from a mentor comes in handy as they encounter a similar situation on a different team. An illuminating metaphor from a book finds its way into a presentation they're delivering. The ideas they've captured begin gravitating toward each other and cross-pollinating.

Fernando is an oncologist at a world-renowned hospital who uses his Second Brain to organize his patient notes. He summarizes the key points from each patient's health history with a special focus on how long they've had the condition, which treatments they've received, and key features of their tumor. Fernando uses his Second Brain to *connect* what he knows from his training and research to the needs of his patients, to be able to treat them more effectively.

Eventually, the third and final way that people use their Second Brain is for creating new things. They realize that they have a lot of knowledge on a subject and decide to turn it into something concrete and shareable. Seeing so much supporting material ready and waiting gives them the courage to put their own ideas out there and have a positive impact on others.

Terrell is a young father of three working a demanding job at a large tech company in Texas. After taking my course, he used his Second Brain to start a YouTube channel where he shares stories and tips on parenting. For example, he's made videos on how to travel internationally with kids and how to request paternity leave, and shared clips from weekend trips he's taken with his family.

Being able to keep track of all the video ideas and production details outside his head has been crucial in allowing Terrell to balance his side gig with his full-time career, while still having enough time with his kids. He is using his Second Brain to express himself and *create* the content he wants to see in the world.

Each of these individuals has leveraged technology to remember, connect, and create far more effectively than if they had to do it on their own.[8] They use their Second Brain in ways that complement the current season of their lives. As those seasons change, they will be able to adapt how they use their notes so that they remain relevant and useful.

Introducing The CODE Method: The Four Steps to Remembering What Matters

To guide you in the process of creating your own Second Brain, I've developed a simple, intuitive four-part method called "CODE"—Capture; Organize; Distill; Express.

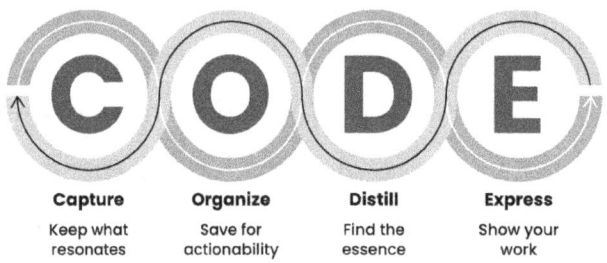

These are the steps not only to build your Second Brain in the first place, but also to work with it going forward. Each of these steps represents a timeless principle found throughout humanity's history, from the earliest cave paintings to the artisan workshops of the Renaissance to the most cutting-edge modern fields. They are flexible and agnostic for any profession, role, or career and for whatever notetaking methods and platforms you prefer. I'm even willing to bet that you're doing them already in some form, whether you realize it or not.

CODE is a map for navigating the endless streams of information we are now faced with every day. It is a modern approach to creating a commonplace book, adapted to the needs of the Information Age.

The same way we have a genetic code that determines our height and eye color, we also have a *creative* code that is hardwired into our imagination. It shapes how we think and how we interact with the world. It is mirrored in the *software* code that runs the apps we use to handle information. It's also been a *secret* code for most of history—now it's finally time to reveal how it works.[III]

Let's preview each of the four steps of the CODE Method—Capture, Organize, Distill, and Express—and then we'll dive into the details in the following chapters.

Capture: Keep What Resonates

Every time we turn on our smartphone or computer, we are immediately immersed in the flow of juicy content they present. Much of this information is useful and interesting—how-to articles that could make us more productive, podcasts with experts sharing hard-won lessons, or inspiring photos of travel destinations we might want to visit.

Here's the problem: we can't consume every bit of this information stream. We will quickly be exhausted and overwhelmed if we try. We need to adopt the perspective of a curator, stepping back from the raging river and starting to make intentional decisions about what information we want to fill our minds.

Like a scientist capturing only the rarest butterflies to take back to the lab, our goal should be to "capture" only the ideas and insights we think are truly noteworthy. Content tends to pile up all around us even without our involvement. There are probably emails filling your inbox, updates popping up in your social media feeds, and notifications proliferating on your smartphone as you're reading this.

It's already there, but we tend to capture it haphazardly, at best. You might email yourself a quick note, brainstorm some ideas in a document, or highlight quotes in a book you're reading, but that information probably remains disconnected and scattered. The insights you uncovered through serious mental effort remain hidden in forgotten folders or drifting in the cloud.

The solution is to *keep only what resonates* in a trusted place that you control, and to leave the rest aside.

When something resonates, it moves you on an intuitive level. Often, the ideas that resonate are the ones that are most unusual, counterintuitive, interesting, or potentially useful. Don't make it an analytical decision, and don't worry about why exactly it resonates—just look inside for a feeling of pleasure, curiosity, wonder, or excitement, and let that be your signal for when it's time to capture a passage, an image, a quote, or a fact.

By training ourselves to notice when something resonates with us, we can improve not only our ability to take better notes, but also our understanding of ourselves and what makes us tick. It is a way of turning up the volume on our intuition so we can hear the wisdom it offers us.

Adopting the habit of knowledge capture has immediate benefits for our mental health and peace of mind. We can let go of the fear that our memory will fail us at a crucial moment. Instead of jumping at every new headline and notification, we can choose to consume information that adds value to our lives and consciously let go of the rest.

Organize: Save for Actionability

Once you've begun capturing notes with the ideas that resonate with you, you'll eventually feel the need to organize them.

It is tempting to try to create a perfect hierarchy of folders up front to contain every possible note you might ever want to capture. Even if it was possible, that approach would be incredibly time-consuming and require way too much effort, pulling you away from what interests you today.

Most people tend to organize information by subject, like the Dewey decimal system you've probably seen at the library. For example, you might find a book under a broad subject category like "Architecture," "Business," "History," or "Geology."

When it comes to digital notes, we can use much easier and lighter ways of organizing. Because our priorities and goals can change at a moment's notice, and probably will, we want to avoid organizing methods that are overly rigid and prescriptive. The best way to organize your notes is to *organize for action*, according to the active projects you are working on right now. Consider new information in terms of its *utility*, asking, "How is this going to help me move forward one of my current projects?"

Surprisingly, when you focus on taking action, the vast amount of information out there gets radically streamlined and simplified. There are relatively few things that are actionable and relevant at any given time, which means you have a clear filter for ignoring everything else.

Organizing for action gives you a sense of tremendous clarity, because you know that everything you're keeping actually has a purpose. You know that it aligns with your goals and priorities. Instead of organizing being an obstacle to your productivity, it becomes a contributor to it.

Distill: Find the Essence

Once you start capturing ideas in a central place and organizing them for action, you'll inevitably begin to notice patterns and connections between them.

An article you read about gardening will give you an insight into growing your customer base. An offhand testimonial from a client will give you the idea to create a web page with all your client testimonials. A business card will remind you of a fascinating conversation you had with someone who you could reach out to for coffee.

The human mind is like a sizzling-hot frying pan of associations—throw a handful of seeds in there and they'll explode into new ideas like popcorn. Every note is the seed of an idea, reminding you of what you already know and already think about a topic.

There is a powerful way to facilitate and speed up this process of rapid association: *distill your notes down to their essence*.

Every idea has an "essence": the heart and soul of what it is trying to communicate. It might take hundreds of pages and thousands of words to fully explain a complex insight, but there is always a way to convey the core message in just a sentence or two.

Einstein famously summarized his revolutionary new theory of physics with the equation $E=mc^2$. If he can distill his thinking into such an elegant equation, you can surely summarize the main points of any article, book, video, or presentation so that the main point is easy to identify.

Why is it so important to be able to easily find the main point of a note? Because in the midst of a busy workday, you won't have time to review ten pages of notes on a book you read last year—you need to be able to quickly find just the main takeaways.

If you've highlighted those takeaways in the flow of the reading you were already doing anyway, you'll be able to remind yourself of what the book contains without having to spend hours rereading it.

Every time you take a note, ask yourself, "How can I make this as useful as possible for my future self?" That question will lead you to annotate the words and phrases that explain why you saved a note, what you were thinking, and what exactly caught your attention.

Your notes will be useless if you can't decipher them in the future, or if they're so long that you don't even try. Think of yourself not just as a *taker* of notes, but as a *giver* of notes—you are giving your future self the gift of knowledge that is easy to find and understand.

Express: Show Your Work

All the previous steps—capturing, organizing, and distilling—are geared toward one ultimate purpose: sharing your own ideas, your own story, and your own knowledge with others.

What is the point of knowledge if it doesn't help anyone or produce anything? [IV] Whether your goal is to lose weight, get a promotion at work, start a side business, or strengthen your local community, personal knowledge management exists to support taking action—anything else is a distraction.

A common challenge for people who are curious and love to learn is that we can fall into the habit of continuously force-feeding ourselves more and more information, but never actually take the next step and apply it. We compile tons of research, but never put forward our own proposal. We gather untold business case studies, but never pitch one potential client. We study every piece of relationship advice available, but never ask anyone out on a date.

It's so easy to endlessly delay and postpone the experiences that would so enrich our lives. We think we're not ready. We fear we're not prepared. We cannot stand the thought that there is one little piece of information we're missing that, if we had it, would make all the difference.

I'm here to tell you that that is no way to live your life. Information becomes *knowledge*—personal, embodied, verified—only when we put it to use. You gain confidence in what you know only when you know that it works. Until you do, it's just a theory.

This is why I recommend you shift as much of your time and effort as possible from consuming to creating.[V] We all naturally have a desire to create—to bring to life something good, true, or beautiful.[2] It's a part of our essential nature. Creating new things is not only one of the most deeply fulfilling things we can do, it can also have a positive impact on others—by inspiring, entertaining, or educating them.

What should you create?

It depends on your skills, interests, and personality. If you are highly analytical, you could evaluate the many options for camping gear and create a list of recommended products to share with your friends. If you like to teach, you could record your favorite dessert recipe and post it on social media or a blog. If you care about a local cause such as public parks, you could create a plan to lobby the city council for more funding.

All these actions—evaluate, share, teach, record, post, and lobby[VI]—are synonyms for the act of expression. They all draw on outside sources for raw material, they all involve a practical process of refinement over time, and they all end up making an impact on someone or something that matters to you.

Information is always in flux, and it is always a work in progress. Since nothing is ever truly final, there is no need to wait to get started. You can publish a simple website now, and slowly add additional pages over time. You can send out a draft of a piece of writing now and make revisions later when you have more time. The sooner you begin, the sooner you start on the path of improvement.

I've introduced a lot of new concepts and terms, and I know at this point it can seem a little overwhelming. It may feel like you have to learn and do a lot of new things to be able to build a Second Brain.

Here's the surprising truth: you are already doing most of the work required.

You are already learning new things—you couldn't stop if you wanted to. You already consume interesting ideas—note the dozens of tabs you likely have open in your browser. You already put in so much effort to keep track of all the information you need for your studies, your job, or your business. All you need is a slightly more intentional, more deliberate way to manage that information, plus a few practical habits to ensure it gets done.

In Part Two, I will show you how to use the CODE steps to radically expand your memory, intelligence, and creativity. For each step, I'll share a set of practical techniques that you can implement today that will begin to yield benefits tomorrow. Techniques that don't require any advanced technology—only the everyday devices and apps you have in your pocket and on your desk right now.

I. Many people who follow the CODE methodology continue to use paper for their notes. Many even find that they take *more* notes on paper once they have a way of capturing those notes digitally and saving them in a secure place. It's not black-or-white. It's about choosing the right tool for the job. This book is focused primarily on the potential of digital notes.

II. Most notes apps provide ways of exporting your notes in standard formats that can then be imported into other apps. I personally have switched platforms twice (from Microsoft Word to Google Docs, and then later to Evernote) and expect to switch to new platforms regularly in the future as technology advances.

III. In a wonderful coincidence, recent research by neurophysiologists May-Britt Moser and Edvard Moser at the Norwegian University of Science and Technology indicates that the human brain remembers information using a "grid code"—a part of the brain involved in spatial reasoning. They speculate that "the grid code could therefore be some sort of metric or coordinate system" that can "uniquely and efficiently represent a lot of information."

IV. The word "productivity" has the same origin as the Latin verb *producere*, which means "to produce." Which means that at the end of the day, if you can't point to some kind of output or result you've produced, it's questionable whether you've been productive at all.

V. The consumerist attitude toward information—that more is better, that we never have enough, and that what we already have isn't good enough—is at the heart of many people's dissatisfaction with how they spend their time online. Instead of trying to find "the best" content, I recommend instead switching your focus to making things, which is far more satisfying.

VI. Other synonyms for expression include publish, speak, present, perform, produce, write, draw, interpret, critique, or translate.

PART TWO

The Method
The Four Steps of CODE

Chapter 4

Capture—Keep What Resonates

> Everything not saved will be lost.
> —Nintendo "Quit Screen" message

Information is food for the brain. It's no accident that we call new ideas "food for thought."

It's clear that we need food and water to survive. What may not be so clear is that we also need information to live: to understand and adapt to our environment; to maintain relationships and cooperate with others; and to make wise decisions that further our interests.

Information isn't a luxury—it is the very basis of our survival.

Just as with the food we put into our bodies, it is our responsibility and right to choose our information diet. It's up to us to decide what information is good for us, what we want more of and less of, and ultimately, what we do with it. You are what you consume, and that applies just as much to information as to nutrition.

A Second Brain gives us a way to filter the information stream and curate only the very best ideas we encounter in a private, trusted place. Think of it as planting your own "knowledge garden" where you are free to cultivate your ideas and develop your own thinking away from the deafening noise of other people's opinions.

A garden is only as good as its seeds, so we want to start by seeding our knowledge garden with only the most interesting, insightful, useful ideas we can find.

You may already consume a lot of content from many different sources, but perhaps never put much thought into what you do with it afterward. Maybe you are already a diligent organizer, but you've fallen into a habit of "digital hoarding" that doesn't end up enriching your life. Or, if this is all completely new to you, you may be starting at square one.

No matter your situation, let's start at the very beginning—how to use the first step of CODE to begin building your own private collection of knowledge.

Building a Private Collection of Knowledge

Taylor Swift is an icon of modern pop and country music and one of the best-selling music artists in history. Her nine chart-topping albums have sold over two hundred million copies worldwide and earned her a long list of awards, including eleven Grammy Awards. Not only does she appear in lists of the greatest singer-songwriters of all time, her influence has transcended music and placed her on such lists as the Time 100 and Forbes Celebrity 100.[1]

Over the course of her career, Swift has released five documentary films revealing her creative songwriting process. In all of them, she can be found with her head buried in her phone. As she says: "I disappear into my phone because my phone is where I keep my notes and my phone is where I'm editing."[2] In her notes she can write down (and reread, edit, and riff off) any snippet of lyrics or melodic hook that flickers through her mind. She can take her notes everywhere, access them from anywhere, and send them within seconds to a wide network of producers and collaborators using the same device. Any feedback they send back can go right into her notes as well.

In an interview about how she wrote the smash hit "Blank Space,"[3] Swift says, "I'll be going about my daily life and I'll think, 'Wow, so we only have two real options in relationships—it's going to be forever or it's going to go down in flames,' so I'll jot that down in my notes… I'll come up with a line that I think is clever like 'Darling I'm a nightmare dressed like a daydream' and I just pick them and put them where they fit and construct the bridge out of more lines that come up within the last couple of years… 'Blank Space' was the culmination of all my best ones one after the other."

For Swift, writing songs is not a discrete activity that she can do only at certain times and in certain places. It is a side effect of the way her mind works, spinning off new metaphors and turns of phrase at the most unexpected times: "I'm inspired to write songs at any time of day, when I'm going through something or when the dust has settled and I'm over it. It can be anything. I'll just be doing dishes or something, or in the middle of an interview, and I could get an idea in my head that just kind of sticks out as, 'That could be a hook, that could be a pre-chorus, that's a first line.'" She goes on to explain why it's so important to her to capture those fleeting thoughts right as they appear: "I kind of have to capitalize on the excitement of me getting that idea and see it all the way through or else I'll leave it behind and assume it wasn't good enough."

Even after all her success, even Taylor Swift needs a system to carry her ideas from inception all the way to completion. By integrating her notetaking with her daily life, she's able to use language and analogies that are rooted in everyday feelings and experiences, forging a powerful connection with her fans who call themselves "Swifties." Listening to her albums is like following Swift on a journey of self-discovery, each album chronicling what she was experiencing and who she was becoming in each chapter of her life.

This story sheds light on how even the world's most successful and prolific creatives need support systems to pursue their craft. It's not a matter of having enough raw talent. Talent needs to be channeled and developed in order to become something more than a momentary spark. Actor and comedian Jerry Seinfeld, arguably the most influential comedian of his generation, wrote in his book *Is This Anything?*:

> *Whenever I came up with a funny bit, whether it happened on a stage, in a conversation or working it out on my preferred canvas, the big yellow legal pad, I kept it in one of those old-school accordion folders... A lot of people I've talked to seemed surprised that I've kept all these notes. I don't understand why they think that. I don't understand why I've kept anything else. What could possibly be of more value?*

Think about your favorite athlete, musician, or actor. Behind the scenes of their public persona, there is a process they follow for regularly turning new ideas into creative output. The same goes for inventors, engineers, and effective leaders.

Innovation and impact don't happen by accident or chance. Creativity depends on a *creative process*.

Creating a Knowledge Bank: How to Generate Compounding Interest from Your Thoughts

In Chapter 2, we looked at the history of commonplace books, kept by intellectuals and writers in previous centuries. For them, the purpose of information was clear: to inform their writing, speaking, and conversation. Knowing how they were going to be putting ideas to use gave them a powerful lens for seeing which ones were worth the trouble of writing down.

This practice continues among creatives today. Songwriters are known for compiling "hook books" full of lyrics and musical riffs they may want to use in future songs. Software engineers build "code libraries" so useful bits of code are easy to access. Lawyers keep "case files" with details from past cases they might want to refer to in the future. Marketers and advertisers maintain "swipe files" with examples of compelling ads they might want to draw from.

The challenge for the rest of us is how to apply this same lens to the work we do every day. What kinds of information are worth preserving when we don't know exactly how we'll be putting it to use? Our world changes much faster than in previous eras, and most of us don't have a single creative medium we work in. How can we decide what to save when we have no idea what the future holds?

To answer that question, we have to radically expand our definition of "knowledge."

Knowledge isn't just wise quotations from long-dead Greek philosophers in white togas. It's not just the teachings found within thick textbooks written by academics with advanced degrees. In the digital world we live in, knowledge most often shows up as "content"—snippets of text, screenshots, bookmarked articles, podcasts, or other kinds of media. This includes the content you gather from outside sources but also the content you create as you compose emails, draw up project plans, brainstorm ideas, and journal your own thoughts.

These aren't just random artifacts with no value—they are "knowledge assets" that crystallize what you know in concrete form.[I]

Knowledge isn't always something "out there" that you have to go out and find. It's everywhere, all around you: buried in the emails in your inbox, hidden within files in your documents folder, and waiting on cloud drives. Knowledge capture is about mining the richness of the reading you're already doing and the life you're already living.

Sometimes these assets are quite mundane—the agenda from last year's financial planning off-site repurposed for next year's meeting. At other times that knowledge is lofty and grand—your in-depth notes from a book on history that could change how you think about the world. Or anything in between. A knowledge asset is anything that can be used in the future to solve a problem, save time, illuminate a concept, or learn from past experience.

Knowledge assets can come from either the external world or your inner thoughts. External knowledge could include:

- **Highlights:** Insightful passages from books or articles you read.

- **Quotes:** Memorable passages from podcasts or audiobooks you listen to.

- **Bookmarks and favorites:** Links to interesting content you find on the web or favorited social media posts.

- **Voice memos:** Clips recorded on your mobile device as "notes to self."

- **Meeting notes:** Notes you take about what was discussed during meetings or phone calls.

- **Images:** Photos or other images that you find inspiring or interesting.

- **Takeaways:** Lessons from courses, conferences, or presentations you've attended.

Look around you and notice that you already have many of these. It may be disorganized, spread around in different places, and saved in different formats, but it's there. Just notice that you've already spent the effort to create or acquire it. All you need to do is gather it up and plant it as the first seeds in your knowledge garden. Soon I'll show you how to do that.

As you start collecting this material from the outer world, it often sparks new ideas and realizations in your inner world. You can capture those thoughts too! They could include:

- **Stories:** Your favorite anecdotes, whether they happened to you or someone else.

- **Insights:** The small (and big) realizations you have.

- **Memories:** Experiences from your life that you don't want to forget.

- **Reflections:** Personal thoughts and lessons written in a journal or diary.

- **Musings:** Random "shower ideas" that pop into your head.

The meaning of a thought, insight, or memory often isn't immediately clear. We need to write them down, revisit them, and view them from a different perspective in order to digest what they mean to us. It is exceedingly difficult to do that within the confines of our heads. We need an external medium in which to see our ideas from another vantage point, and writing things down is the most effective and convenient one ever invented.

Perhaps you have some hesitation about writing down such personal thoughts in a piece of software rather than a private journal. While it's always up to you what you choose to note down, remember that your Second Brain is private too. You can share certain notes if you want to, but by default everything inside is for your eyes only.

For now, choose the two to three kinds of content from the two lists above that you already have the most of and already value. Some people favor inner sources of knowledge, some people are biased toward the outer world, but most people are somewhere in between. While you can eventually learn to capture from dozens of different sources, it's important to start small and get your feet wet before diving into the deep end.

What *Not* to Keep

The examples I've shared may seem so expansive that you're wondering if there is anything you *shouldn't* keep in your Second Brain. In my experience, there are four kinds of content that aren't well suited to notes apps:

- **Is this sensitive information you'd like to keep secure?** The content you save in your notes is easily accessible from any device, which is great for accessibility but not for security. Information like tax records, government documents, passwords, and health records shouldn't be saved in your notes.

- **Is this a special format or file type better handled by a dedicated app?** Although you could save specialized files such as Photoshop files or video footage in your notes, you'll need a specialized app to open them anyway, so there's no benefit to keeping them in your notes.

- **Is this a very large file?** Notes apps are made for short, lightweight bits of text and images, and their performance will often be severely hampered if you try to save large files in them.

- **Will it need to be collaboratively edited?** Notes apps are perfectly suited for individual, private use, which makes them less than ideal for collaboration. You can share individual notes or even groups of notes with others, but if you need multiple people to be able to collaboratively edit a document in real time, then you'll need to use a different platform.

Twelve Favorite Problems: A Nobel Prize Winner's Approach to Capturing

With the abundance of content all around us, it can be hard to know exactly what is worth preserving. I use an insightful exercise to help people make this decision easier. I call it "Twelve Favorite Problems," inspired by Nobel Prize–winning physicist Richard Feynman.

Feynman was known for his wide-ranging, eclectic tastes. As a child he already showed a talent for engineering, once building a functioning home alarm system out of spare parts while his parents were out running errands. During his colorful lifetime Feynman spent time in Brazil teaching physics, learned to play the bongo and the conga drums well enough to perform with orchestras, and enthusiastically traveled around the world exploring other cultures.

Of course, Feynman is best known for his groundbreaking discoveries in theoretical physics and quantum mechanics, for which he received the Nobel Prize in 1965. In his spare time, he also played a pivotal role on the commission that investigated the *Challenger* space shuttle disaster and published half a dozen books.

How could one person make so many contributions across so many areas? How did he have the time to lead such a full and interesting life while also becoming one of the most recognized scientists of his generation?

Feynman revealed his strategy in an interview[4]:

You have to keep a dozen of your favorite problems constantly present in your mind, although by and large they will lay in a dormant state. Every time you hear or read a new trick or a new result, test it against each of your twelve problems to see whether it helps. Every once in a while there will be a hit, and people will say, "How did he do it? He must be a genius!"

In other words, Feynman's approach was to maintain a list of a dozen open questions. When a new scientific finding came out, he would test it against each of his questions to see if it shed any new light on the problem. This cross-disciplinary approach allowed him to make connections across seemingly unrelated subjects, while continuing to follow his sense of curiosity.

As told in *Genius: The Life and Science of Richard Feynman* by James Gleick,[5] Feynman once took inspiration for his physics from an accident at dinner:

... he was eating in the student cafeteria when someone tossed a dinner plate into the air—a Cornell cafeteria plate with the university seal imprinted on one rim—and in the instant of its flight he experienced what he long

afterward considered an epiphany. As the plate spun, it wobbled. Because of the insignia he could see that the spin and the wobble were not quite in synchrony. Yet just in that instant it seemed to him—or was it his physicist's intuition?—that the two rotations were related.

After working the problem out on paper, Feynman discovered a 2-to-1 ratio between the plate's wobble and spin, a neat relationship that suggested a deeper underlying principle at work.

When a fellow physicist and mentor asked what the use of such an insight was, Feynman responded: "It doesn't have any importance... I don't care whether a thing has importance. Isn't it fun?" He was following his intuition and curiosity. But it did end up having importance, with his research into the equations underlying rotation informing the work that ultimately led to him receiving the Nobel Prize.

Feynman's approach encouraged him to follow his interests wherever they might lead. He posed questions and constantly scanned for solutions to long-standing problems in his reading, conversations, and everyday life. When he found one, he could make a connection that looked to others like a flash of unparalleled brilliance.

Ask yourself, "What are the questions I've always been interested in?" This could include grand, sweeping questions like "How can we make society fairer and more equitable?" as well as practical ones like "How can I make it a habit to exercise every day?" It might include questions about relationships, such as "How can I have closer relationships with the people I love?" or productivity, like "How can I spend more of my time doing high-value work?"

Here are more examples of favorite problems from my students:

- How do I live less in the past, and more in the present?

- How do I build an investment strategy that is aligned with my mid-term and long-term goals and commitments?

- What does it look like to move from mindless consumption to mindful creation?

- How can I go to bed early instead of watching shows after the kids go to bed?
- How can my industry become more ecologically sustainable while remaining profitable?
- How can I work through the fear I have of taking on more responsibility?
- How can my school provide more resources for students with special needs?
- How do I start reading all the books I already have instead of buying more?
- How can I speed up and relax at the same time?
- How can we make the healthcare system more responsive to people's needs?
- What can I do to make eating healthy easier?
- How can I make decisions with more confidence?

Notice that some of these questions are abstract, while others are concrete. Some express deep longings, while others are more like spontaneous interests. Many are questions about how to live a better life, while a few are focused on how to succeed professionally. The key to this exercise is to make them *open-ended* questions that don't necessarily have a single answer. To find questions that invoke a state of wonder and curiosity about the amazing world we live in.

The power of your favorite problems is that they tend to stay fairly consistent over time. The exact framing of each question may change, but even as we move between projects, jobs, relationships, and careers, our favorite problems tend to follow us across the years. I recommend asking your family or childhood friends what you were obsessed with as a kid. Those very same interests probably still fire your imagination as an adult. Which means any content you collect related to them will likely be relevant far into the future as well.

As a kid, I had a passion for LEGOs, the modular toy blocks beloved by generations of children. My parents noticed that I didn't play with LEGOs like other kids. Instead I spent my time organizing and reorganizing the pieces. I remember being completely captivated by the problem of how to create order out of the chaos of thousands of pieces of every shape and size. I would invent new organizational schemes—by color, by size, by theme—as I became obsessed with the idea that if I could just find the right system, I would finally be able to build my magnum opus—a LEGO spaceship like the ones I saw in the sci-fi movies I loved.

That very same question—How can creativity emerge out of chaos?—still drives me to this day. Only now, it's in the form of organizing digital information instead of LEGOs. Pursuing this question has taught me so many things over the years, across many seasons of my life. The goal isn't to definitively answer the question once and for all, but to use the question as a North Star for my learning.

Take a moment now to write down some of your own favorite problems. Here are my recommendations to guide you:

- Ask people close to you what you were obsessed with as a child (often you'll continue to be fascinated with the same things as an adult).

- Don't worry about coming up with exactly twelve (the exact number doesn't matter, but try to come up with at least a few).

- Don't worry about getting the list perfect (this is just a first pass, and it will always be evolving).

- Phrase them as open-ended questions that could have multiple answers (in contrast to "yes/no" questions with only one answer).

Use your list of favorite problems to make decisions about what to capture: anything potentially relevant to answering them. Use one of the capture tools I recommend later in this chapter, or in the Second Brain Resource Guide at Buildingasecondbrain.com/resources.

Capture Criteria: How to Avoid Keeping Too Much (or Too Little)

Once you have identified the kinds of questions you want your Second Brain to answer, it's time to choose specifically which pieces of information will be most useful.

Imagine you come across a blog post during your web browsing that details how a marketing expert you respect runs her campaigns. You're hooked: this is the kind of material you've been looking for! Finally, the master reveals her secrets!

Your first instinct might be to save the article in its entirety. It's high-quality information, so why not preserve all of it? The problem is, it's an in-depth how-to article that is thousands of words long. Even if you spend the twenty or thirty minutes it would take to consume it now, in the future you'll just have to spend all that time reading it again, since you'll have forgotten most of the details. You also don't want to just bookmark the link and save it to read later, because then you won't know what it contains in the first place!

This is where most people get stuck. They either dive straight into the first piece of content they see, read it voraciously, but quickly forget all the details, or they open dozens of tabs in their web browser and feel a pang of guilt at all those interesting resources they haven't been able to get to.

There is a way out of this situation. It starts with realizing that in any piece of content, *the value is not evenly distributed*. There are always certain parts that are especially interesting, helpful, or valuable to you. When you realize this, the answer is obvious. You can extract only the most salient, relevant, rich material and save it as a succinct note.

Don't save entire chapters of a book—save only select passages. Don't save complete transcripts of interviews—save a few of the best quotes. Don't save entire websites—save a few screenshots of the sections that are most interesting. The best curators are picky about what they allow into their collections, and you should be too. With a notes app, you can always save links back to the original content if you need to review your sources or want to dive deeper into the details in the future.

The biggest pitfall I see people falling into once they begin capturing digital notes is saving too much. If you try to save every piece of material you come across, you run the risk of inundating your future self with tons of irrelevant information. At that point, your Second Brain will be no better than scrolling through social media.

This is why it's so important to take on a Curator's Perspective—that we are the judges, editors, and interpreters of the information we choose to let into our lives. Thinking like a curator means taking charge of your own information stream, instead of just letting it wash over you. The more economical you can be with the material you capture in the first place, the less time and effort your future self will have to spend organizing, distilling, and expressing it.[II]

Here are four criteria I suggest to help you decide exactly which nuggets of knowledge are worth keeping:

Capture Criteria #1: Does It Inspire Me?

Inspiration is one of the most rare and precious experiences in life. It is the essential fuel for doing your best work, yet it's impossible to call up inspiration on demand. You can Google the answer to a question, but you can't Google a feeling.

There is a way to evoke a sense of inspiration more regularly: keep a collection of inspiring quotes, photos, ideas, and stories. Any time you need a break, a new perspective, or a dash of motivation, you can look through it and see what sparks your imagination.

For example, I keep a folder full of customer testimonials I've received over the years. Any time I think what I'm doing doesn't matter or isn't good enough, all I have to do is open up that folder and my perspective is completely shifted.

Capture Criteria #2: Is It Useful?

Carpenters are known for keeping odds and ends in a corner of their workshop—a variety of nails and washers, scraps of lumber cut off from larger planks, and random bits of metal and wood. It costs nothing to keep these "offcuts" around, and surprisingly often they end up being the crucial missing piece in a future project.

Sometimes you come across a piece of information that isn't necessarily inspiring, but you know it might come in handy in the future. A statistic, a reference, a research finding, or a helpful diagram—these are the equivalents of the spare parts a carpenter might keep around their workshop.

For example, I keep a folder full of stock photos, graphics, and drawings I find both online and offline. Any time I need an image for a slide deck, or a web page, or to spark new ideas, I have a plentiful supply of imagery I've already found compelling ready and waiting.

Capture Criteria #3: Is It Personal?

One of the most valuable kinds of information to keep is personal information—your own thoughts, reflections, memories, and mementos. Like the age-old practice of journaling or keeping a diary, we can use notetaking to document our lives and better understand how we became who we are.

No one else has access to the wisdom you've personally gained from a lifetime of conversations, mistakes, victories, and lessons learned. No one else values the small moments of your days quite like you do.

I often save screenshots of text messages sent between my family and friends. The small moments of warmth and humor that take place in these threads are precious to me, since I can't always be with them in person. It takes mere moments, and I love knowing that I'll forever have memories from my conversations with the people closest to me.

Capture Criteria #4: Is It Surprising?

I've often noticed that many of the notes people take are of ideas they already know, already agree with, or could have guessed. We have a natural bias as humans to seek evidence that confirms what we already believe, a well-studied phenomenon known as "confirmation bias."[6]

That isn't what a Second Brain is for. The renowned information theorist Claude Shannon, whose discoveries paved the way for modern technology, had a simple definition for "information": that which surprises you.[7] If you're not surprised, then you already knew it at some level, so why take note of it? Surprise is an excellent barometer for information that doesn't fit neatly into our existing understanding, which means it has the potential to change how we think.

Sometimes you come across an idea that is neither inspiring, personal, nor obviously useful, but there is something surprising about it. You may not be able to put your finger on why, but it conflicts with your existing point of view in a way that makes your brain perk up and pay attention. Those are the ideas you should capture.

Your Second Brain shouldn't be just another way of confirming what you already know. We are already surrounded by algorithms that feed us only what we already believe and social networks that continually reinforce what we already think.

Our ability to capture ideas from anywhere takes us in a different direction: By saving ideas that may contradict each other and don't necessarily support what we already believe, we can train ourselves to take in information from different sources instead of immediately jumping to conclusions. By playing with ideas—bending and stretching and remixing them—we become less attached to the way they were originally presented and can borrow certain aspects or elements to use in our own work.

If what you're capturing doesn't change your mind, then what's the point?

Ultimately, Capture What Resonates

I've given you specific criteria to help you decide what is worth capturing, but if you take away one thing from this chapter, it should be to keep what resonates.

Here's why: making decisions analytically, with a checklist, is taxing and stressful. It is the kind of thinking that demands the most energy. When you use up too much energy taking notes, you have little left over for the subsequent steps that add far more value: making connections, imagining possibilities, formulating theories, and creating new ideas of your own. Not to mention, if you make reading and learning into unpleasant experiences, over time you're going to find yourself doing less and less of them. The secret to making reading a habit is to make it effortless and enjoyable.

As you consume a piece of content, listen for an internal feeling of being moved or surprised by the idea you're taking in. This special feeling of "resonance"—like an echo in your soul—is your intuition telling you that something is literally "noteworthy." You don't need to figure out exactly why it resonates. Just look for the signs: your eyes might widen slightly, your heart may skip a beat, your throat may go slightly dry, and your sense of time might subtly slow down as the world around you fades away. These are clues that it's time to hit "save."

We know from neuroscientific research that "emotions organize—rather than disrupt—rational thinking."[8] When something resonates with us, it is our emotion-based, intuitive mind telling us it is interesting before our logical mind can explain why. I often find that a piece of content resonates with me in ways I can't fully explain in the moment, and its true potential only becomes clear later on.

There's scientific evidence that our intuition knows what it's doing. From the book *Designing for Behavior Change*:[9]

> *Participants in a famous study were given four biased decks of cards—some that would win them money, and some that would cause them to lose. When they started the game, they didn't know that the decks were biased. As they played the game, though, people's bodies started showing signs of physical "stress" when their conscious minds were about to use a money-losing deck. The stress was an automatic response that occurred because the intuitive mind realized something was wrong—long before the conscious mind realized anything was amiss.*

The authors' conclusion: "Our intuitive mind learns, and responds, even without our conscious awareness."

If you ignore that inner voice of intuition, over time it will slowly quiet down and fade away. If you practice listening to what it is telling you, the inner voice will grow stronger. You'll start to hear it in all kinds of situations. It will guide you in what choices to make and which opportunities to pursue. It will warn you away from people and situations that aren't right for you. It will speak up and take a stand for your convictions even when you're afraid.

I can't think of anything more important for your creative life—and your life in general—than learning to listen to the voice of intuition inside. It is the source of your imagination, your confidence, and your spontaneity. You can intentionally train yourself to hear that voice of intuition every day by taking note of what it tells you.

Besides capturing what personally resonates with you, there are a couple other kinds of details that are generally useful to save in your notes. It's a good idea to capture key information about the source of a note, such as the original web page address, the title of the piece, the author or publisher, and the date it was published.[III] Many capture tools are even able to identify and save this information automatically. Also, it's often helpful to capture chapter titles, headings, and bullet-point lists, since they add structure to your notes and represent distillation already performed by the author on your behalf.

Beyond Your Notetaking App: Choosing Capture Tools

Now that you know what kinds of material to save in your Second Brain, it's time to get into the nitty-gritty: How does capturing work exactly?

Let's say while reading that in-depth marketing article, you decide a specific piece of advice is highly relevant to your own plans. Most notetaking apps (introduced in Chapter 2 and covered in detail in the Second Brain Resource Guide at Buildingasecondbrain.com/resources) have built-in features that allow you to capture excerpts from outside sources, and you can always simply cut and paste text directly into a new note. There is also an array of more specialized "capture tools" that are designed to make capturing content in digital form easy and even fun.

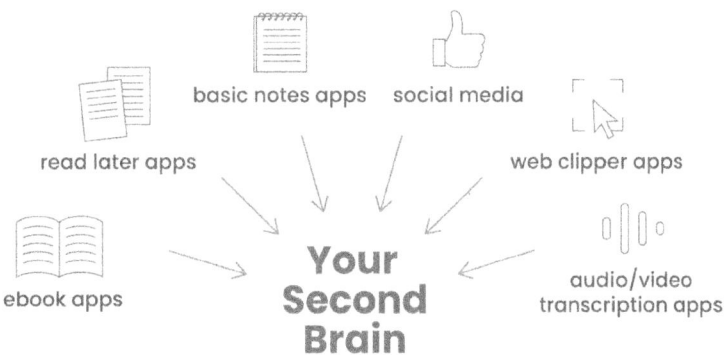

The most common options include:

- Ebook apps, which often allow you to export your highlights or annotations all at once.

- Read later apps that allow you to bookmark content you find online for later reading (or in the case of podcasts or videos, listening or watching).

- Basic notes apps that often come preinstalled on mobile devices and are designed for easily capturing short snippets of text.

- Social media apps, which usually allow you to "favorite" content and export it to a notes app.

- Web clippers, which allow you to save parts of web pages (often included as a built-in feature of notes apps).

- Audio/voice transcription apps that create text transcripts from spoken words.

- Other third-party services, integrations, and plug-ins that automate the process of exporting content from one app to another.

Some of these tools are free, and others charge a small fee. Some are completely automatic, working silently in the background (for example, to automatically sync your ebook highlights with a notes app), while others require a bit of manual effort (such as taking photos of paper notebooks to save them digitally).[IV] But in any case, the act of capture takes only moments—to hit share, export, or save—

and voilà, you've preserved the best parts of whatever you're consuming in your Second Brain.

Make no mistake: you will continue to use many kinds of software to manage information—such as computer folders, cloud storage drives, and various platforms for sharing and collaborating on documents. Think of your capture tools as your extended nervous system, reaching out into the world to allow you to sense your surroundings. No matter how many different kinds of software you use, don't leave all the knowledge they contain scattered across dozens of places you'll never think to look. Make sure your best findings get routed back to your notes app where you can put them all together and act on them.

Here are some of the most popular ways of using capture tools to save content you come across:

- **Capturing passages from ebooks:** Most ebook apps make it very easy to highlight passages as you read. On Amazon Kindle, you can simply drag your finger across a sentence or paragraph you like to add a highlight. Then use the share menu to export all your highlights from the entire book all at once straight to your digital notes. You can also add comments right alongside the text as you read, which will help you remember what you found interesting about a passage.

- **Capturing excerpts from online articles or web pages:** When you come across an online article or blog post you want to read, save it to a "read later" app, which is like a digital magazine rack of everything you want to read (or watch or listen to) at some point. Whenever you have some free time (such as on breaks or in the evening after work), scroll through the articles you've saved and pick one to read. You can make highlights, just as with ebooks, and they can also be automatically exported to your notes app using a third-party platform.

- **Capturing quotes from podcasts:** Many podcast player apps allow you to bookmark or "clip" segments of episodes as you're listening to them. Some of them will even transcribe the audio into text, so you can export and search it within your notes.

- **Capturing voice memos:** Use a voice memo app that allows you to press a button, speak directly into your smartphone, and have every word transcribed into text and exported to your notes.

- **Capturing parts of YouTube videos:** This is a little-known feature, but almost every YouTube video is accompanied by an automatically generated transcript. Just click the "Open transcript" button and a window will open. From there, you can copy and paste excerpts to your notes.

- **Capturing excerpts from emails:** Most popular notes apps include a feature that allows you to forward any email to a special address, and the full text of that email (including any attachments) will be added to your notes.

- **Capturing content from other apps:** You might edit photos in a photos app, or make sketches in a drawing app, or like posts in a social media app. As long as that app has a "share" button or allows for copy-and-paste, you can save whatever you've created directly to your notes for safekeeping.

The Surprising Benefits of Externalizing Our Thoughts

Often ideas occur to us at the most random times—during our commute, while watching TV, when we're playing with our kids, or in the shower.

Your Second Brain gives you a place to corral the jumble of thoughts tumbling through your head and park them in a waiting area for safekeeping. Not only does this allow you to preserve them for the long term; there are multiple other profound benefits that come from the simple act of writing something down.

First, you are much more likely to remember information you've written down in your own words. Known as the "Generation Effect,"[10] researchers have found that when people actively generate a series of words, such as by speaking or writing, more parts of their brain are activated when compared to simply reading the same words. Writing things down is a way of "rehearsing" those ideas, like practicing a dance routine or shooting hoops, which makes them far more likely to stick.

Enhancing our memory is just the beginning. When you express an idea in writing, it's not just a matter of transferring the exact contents of your mind into paper or digital form. Writing creates new knowledge that wasn't there before. Each word you write triggers mental cascades and internal associations, leading to further ideas, all of which can come tumbling out onto the page or screen.[v]

Thinking doesn't just produce writing; writing also enriches thinking.

There is even significant evidence that expressing our thoughts in writing can lead to benefits for our health and well-being.[11] One of the most cited psychology papers of the 1990s found that "translating emotional events into words leads to profound social, psychological, and neural changes."

In a wide range of controlled studies, writing about one's inner experiences led to a drop in visits to the doctor, improved immune systems, and reductions in distress. Students who wrote about emotional topics showed improvements in their grades, professionals who had been laid off found new jobs more quickly, and staff members were absent from work at lower rates. The most amazing thing about these findings is that they didn't rely on input from others. No one had to read or respond to what these people wrote down—the benefits came just from the act of writing.

Perhaps the most immediate benefit of capturing content outside our heads is that we escape what I call the "reactivity loop"—the hamster wheel of urgency, outrage, and sensationalism that characterizes so much of the Internet. The moment you first encounter an idea is the worst time to decide what it means. You need to set it aside and gain some objectivity.

With a Second Brain as a shield against the media storm, we no longer have to react to each idea immediately, or risk losing it forever. We can set things aside and get to them later when we are calmer and more grounded. We can take our time slowly absorbing new information and integrating it into our thinking, free of the pressing demands of the moment. I'm always amazed that when I revisit the items I've previously saved to read later, many of them that seemed so important at the time are clearly trivial and unneeded.

Notetaking is the easiest and simplest way of externalizing our thinking. It requires no special skill, is private by default, and can be performed anytime and anywhere. Once our thoughts are outside our head, we can examine them, play with them, and make them better. It's like a shortcut to realizing the full potential of the thoughts flowing through our minds.

Your Turn: What Would This Look Like If It Was Easy?

I've introduced a lot of ideas in this chapter, and I know it's a lot to absorb. There are so many ways to capture knowledge, but when you're just getting started all those options can just feel overwhelming.

I want to give you an open question that will help guide you as you embark on this journey: What would capturing ideas look like if it was easy?

Think about what you would want to capture more of (or less of). How would that feel? What kinds of content are already familiar enough that it would be easy to begin saving them now? What would capturing look like today or this week? On average I capture just two notes per day—what are two ideas, insights, observations, perspectives, or lessons you've encountered today that you could write down right now?

It's important to keep capturing relatively effortless because it is only the first step. You need to do it enough that it becomes second nature, while conserving your time and energy for the later steps when the value of the ideas you've found can be fully unleashed.

Capture isn't about doing more. It's about taking notes on the experiences you're already having. It's about squeezing more juice out of the fruit of life, savoring every moment to the fullest by paying closer attention to the details.

Don't worry about whether you're capturing "correctly." There's no right way to do this, and therefore, no wrong way. The only way to know whether you're getting the good stuff is to try putting it to use in real life. We'll get to that soon, but in the meantime, try out a couple of digital notes apps and capture tools to see which ones fit your style. Don't forget the resource guide I've put together to help you make your choice.

If at any point you feel stuck or overwhelmed, step back and remember that nothing is permanent in the digital world. Digital content is endlessly malleable, so you don't have to commit to any decision forever. While every step of CODE complements the others, you can also use them one at a time. Start with the parts that resonate with you and expand from there as your confidence grows.

In the next chapter, I'll tell you what to do with the knowledge assets you've gathered in your Second Brain.

I. MIT economist César Hidalgo in his book *Why Information Grows* describes how physical products, which he calls "crystals of imagination," allow us to turn what we know into concrete objects that other people can access: "Crystallizing our thoughts into tangible and digital objects is what allows us to share our thoughts with others." And elsewhere: "Our ability to crystallize imagination… gives us access to the practical uses of the knowledge and knowhow residing in the nervous systems of other people."

II. If you're looking for a more precise answer of how much content to capture in your notes, I recommend no more than 10 percent of the original source, at most. Any more than that, and it will be too difficult to wade through all the material in the future. Conveniently, 10 percent also happens to be the limit that most ebooks allow you to export as highlights.

III. Even if the original web page disappears, you can often use this information to locate an archived version using the Wayback Machine, a project of the Internet Archive that preserves a record of websites: https://archive.org/web/.

IV. The software landscape is constantly changing, so I've created a resource guide with my continually updated recommendations of the best capture tools, both free and paid, and for a variety of devices and operating systems. You can find it at buildingasecondbrain.com/resources.

V. This is called "detachment gain," as explained in *The Detachment Gain: The Advantage of Thinking Out Loud* by Daniel Reisberg, and refers to the "functional advantage to putting thoughts into externalized forms" such as speaking or writing, leading to the "possibility of new discoveries that might not have been obtained in any other fashion." If you've ever had to write out a word to remember how it's spelled, you've experienced this.

Chapter 5

Organize—Save for Actionability

Be regular and orderly in your life so that you may be violent and original in your work.
—Gustave Flaubert, French novelist

Twyla Tharp is one of the most celebrated, inventive dance choreographers in modern times. Her body of work is made up of more than 160 pieces, including 129 dances, twelve television specials, six major Hollywood movies, four full-length ballets, four Broadway shows, and two figure-skating routines.

Dance might seem like the creative medium that could least benefit from "organizing." It is performed live each time, using primarily the dancers' own bodies, and often seems improvisational and spontaneous. Yet in her book *The Creative Habit*,[1] Tharp revealed that a simple organizing technique lies at the heart of a creative process that has propelled her through an incredibly prolific six-decade career.

Tharp calls her approach "the box." Every time she begins a new project, she takes out a foldable file box and labels it with the name of the project, usually the name of the dance she is choreographing. This initial act gives her a sense of purpose as she begins: "The box makes me feel organized, that I have my act together even when I don't know where I'm going yet. It also represents a commitment. The simple act of writing a project name on the box means I've started work."

Into the box she puts anything and everything related to the project, like a swirling cauldron of creative energy. Any time she finds a new piece of material—such as "notebooks, news clippings, CDs, videotapes of me working alone in my studio, videos of the dancers rehearsing, books and photographs and pieces of art that may have inspired me"—she always knows where to put it. It all goes into the box. Which means that any time she works on that project, she knows exactly where to look—in the box.

In her book, Tharp tells the story of a specific project where the box proved invaluable: a collaboration with the pop rock icon Billy Joel to turn a collection of his songs into a full-length dance performance. It was a bold idea, somewhere between a concert and a musical, but quite distinct from either one. It wasn't clear how the characters in different songs, who weren't written as part of the same story, could be combined into the same narrative.

Even a project as open-ended as this one started the same way as all the others, with her goals: "I believe in starting each project with a stated goal. Sometimes the goal is nothing more than a personal mantra such as 'keep it simple' or 'something perfect' or 'economy' to remind me of what I was thinking at the beginning if and when I lose my way. I write it down on a slip of paper and it's the first thing that goes into the box."

For the collaboration with Joel, Tharp had two goals: The first was to understand and master the role of narrative in dance, a long-standing creative challenge that had captured her curiosity. The second goal was much more practical, but no less motivating: to pay her dancers well. She said, "So I wrote my goals for the project, 'tell a story' and 'make dance pay,' on two blue index cards and watched them float to the bottom of the Joel box… they sit there as I write this, covered by months of research, like an anchor keeping me connected to my original impulse."

After that, every bit of research and every idea potentially relevant to the project went into Tharp's box. Recordings of Billy Joel's music videos, live performances, lectures, photographs, news clippings, song lists, and notes about those songs. She gathered news footage and movies about the Vietnam War, important books from the era, and even material from other boxes, including research from an abandoned project that never made it onto the stage.

The artifacts that Tharp collected weren't just for her own use. They became sparks of inspiration for her team: a pair of earrings and a macramé vest shared with the costume designer; books about psychedelic light events to inspire the lighting designer; photographs from other shows and Joel's childhood home in Long Island to discuss with the production designer.

All this creative raw material eventually filled twelve boxes, but all the collecting and gathering from the outside world doesn't mean that Tharp didn't add her own creativity. For example, she found an elaborate set of notes from an early song of Joel's called "She's Got a Way," which was full of innocence and sweetness. She decided to change its meaning: "In my notes you can see the song morphing into something harsher, eventually becoming two simultaneous sleazy bar scenes, one in Vietnam, the other back home. I felt obliged to run this by Billy, warning him, 'This is going to destroy the song.' He wasn't worried. 'Go for it,' he said."

Twyla Tharp's box gave her several powerful advantages as she set out on her creative journey.

The box gave her the security to venture out and take risks: "a box is like soil to me. It's basic, earthy, elemental. It's home. It's what I can always go back to when I need to regroup and keep my bearings. Knowing that the box is always there gives me the freedom to venture out, be bold, dare to fall flat on my face."

The box gave Tharp a way to put projects on hold and revisit them later: "The box makes me feel connected to a project… I feel this even when I've back-burnered a project: I may have put the box away on a shelf, but I know it's there. The project name on the box in bold black lettering is a constant reminder that I had an idea once and may come back to it very soon."

Finally, it gave her a way to look back on her past victories: "There's one final benefit to the box: It gives you a chance to look back. A lot of people don't appreciate this. When they're done with a project, they're relieved. They're ready for a break and then they want to move forward to the next idea. The box gives you the opportunity to reflect on your performance. Dig down through the boxes archaeologically and you'll see a project's beginnings. This can be instructive. How did you do? Did you get to your goal? Did you improve on it? Did it change along the way? Could you have done it all more efficiently?"

Twyla Tharp's box reveals the true value of a simple container: it is easy to use, easy to understand, easy to create, and easy to maintain. It can be moved from place to place without losing its contents. A container requires no effort to identify, to share with others, and to put in storage when it's no longer needed. We don't need complex, sophisticated systems to be able to produce complex, sophisticated works.

The Cathedral Effect: Designing a Space for Your Ideas

Consider how much time we spend designing and arranging our physical environment.

We buy nice furniture, deliberate for weeks over the color of our walls, and fiddle with the placement of plants and books. We know that the details of lighting, temperature, and the layout of a space dramatically affect how we feel and think.

There's a name for this phenomenon: the Cathedral Effect.[2] Studies have shown that the environment we find ourselves in powerfully shapes our thinking. When we are in a space with high ceilings, for example—think of the lofty architecture of classic churches invoking the grandeur of heaven—we tend to think in more abstract ways. When we're in a room with low ceilings, such as a small workshop, we're more likely to think concretely.

No one questions the importance of having physical spaces that make us feel calm and centered, but when it comes to your digital workspace, it's likely you've spent little time, if any, arranging that space to enhance your productivity or creativity. As knowledge workers we spend many hours every day within digital environments—our computers, smartphones, and the web. Unless you take control of those virtual spaces and shape them to support the kinds of thinking you want to do, every minute spent there will feel taxing and distracting.

Your Second Brain isn't just a tool—it's an environment. It is a garden of knowledge full of familiar, winding pathways, but also secret and secluded corners. Every pathway is a jumping-off point to new ideas and perspectives. Gardens are natural, but they don't happen by accident. They require a caretaker to seed the plants, trim the weeds, and shape the paths winding through them. It's time for us to put more intention into the digital environments where we now spend so many of our waking hours.

Once you've created this environment, you'll know where to go when it's time to execute or create. You won't have to sit down and spend half an hour painstakingly gathering together all the materials you need to get started. Your Second Brain is like a mind cathedral that you can step into any time you want to shut out the world and imagine a world of your own.

The next step in building your Second Brain is to take the morsels of insight you've begun to capture and organize them in a space where you can do your best thinking.

Organizing for Action: Where 99 Percent of Notetakers Get Stuck (And How to Solve It)

As you begin to capture your ideas in a consistent way, you're likely to experience a new sense of excitement about the information flowing all around you.

You'll start to pay closer attention to the books you read, the conversations you have, and the interviews you listen to, knowing that any interesting idea you encounter can be reliably saved and utilized. You no longer have to hope that you'll remember your best ideas—you can ensure that you will.

Soon, however, you'll run head on into a new problem: what to do with all this valuable material you've gathered. The more diligently you collect it, the bigger this problem will be! Capturing notes without an effective way to organize and retrieve them only leads to more overwhelm.

I spent years trying different ways of solving the problem of how to organize my digital life. I tried techniques borrowed from organizing physical spaces, every kind of specially formatted notebook, and even the Dewey decimal system used in libraries. I tried organizing my files by date, by subject, by kind, and countless other elaborate schemes, but every method I tried soon failed.

The problem was that none of these systems was integrated into my daily life. They always required me to follow a series of elaborate rules that took time away from my other priorities, which meant they would quickly become outdated and obsolete. Every time I fell off the organizing wagon, I reverted to dropping all my notes and files into a folder for whichever project I was currently focused on. This ensured that at least I had exactly what I needed for my current work immediately on hand—no tagging, filing, or keywords needed.

Then one day I had a realization: Why didn't I just organize my files that way all the time? If organizing by project was the most natural way to manage information with minimal effort, why not make it the default?

That is what I did, and to my surprise, it worked. Over time, I refined, simplified, and tested this action-based approach with thousands of students and followers. I eventually named this organizing system PARA,[I] which stands for the four main categories of information in our lives: Projects, Areas, Resources, and Archives. These four categories are universal, encompassing *any* kind of information, from *any* source, in *any* format, for *any* purpose.[II]

PARA can handle it all, regardless of your profession or field, for one reason: it organizes information based on *how actionable it is*, not *what kind of information it is*. The project becomes the main unit of organization for your digital files. Instead of having to sort your notes according to a complex hierarchy of topics and subtopics, you have to answer only one simple question: "In which project will this be most useful?" It assumes only that you are currently working on a certain set of projects, and that your information should be organized to support them.

For example, say you come across a useful article on how to become more resilient and capture it in your notes. You're sure that this information will come in handy one day, but how do you know where to put those notes in the meantime? How will you remember where to look the next time you need them? This can quickly become an anxiety-provoking decision because of the risk of making the wrong choice.

Most people would save this note by subject in a folder called "Psychology." That seems like a perfectly logical choice. Here's the problem: the subject of "Psychology" is far too broad to be useful. Imagine your future self a few weeks or months from now. In the middle of your workday, how much time will you have to search through all your notes on such a broad subject? There might be notes from many dozens of articles, books, and other resources in there, much of which won't be actionable at all. It would take hours just to figure out what you have.

There's another way. I will show you how to take the notes you've captured and save them according to a practical use case. By taking that small extra step of putting a note into a folder (or tagging it[III]) for a specific project, such as a psychology paper you're writing or a presentation you're preparing, you'll encounter that idea right at the moment it's most relevant. Not a moment before, and not a moment after.

If there isn't a current project that your note would be useful for, we have a couple of other options of where to put it, including dedicated places for each of the main "areas" of your life that you are responsible for, and "resources," which is like a personal library of references, facts, and inspiration. Over time, as you complete your projects, master new skills, and progress toward your goals, you'll discover that some notes and resources are no longer actionable. I'll show you how to move them to your "archives" to keep them out of sight but within easy reach.

These four categories—Projects, Areas, Resources, and Archives—make up the four categories of PARA. We'll explore each of them soon.

Instead of requiring tons of time meticulously organizing your digital world, PARA guides you in quickly sorting your ideas according to what really matters: your goals.

One of the biggest temptations with organizing is to get too perfectionistic, treating the process of organizing as an end in itself. There is something inherently satisfying about order, and it's easy to stop there instead of going on to develop and share our knowledge. We need to always be wary of accumulating so much information that we spend all our time managing it, instead of putting it to use in the outside world.

Instead of inventing a completely different organizational scheme for every place you store information, which creates a tremendous amount of friction navigating the inconsistencies between them, PARA can be used everywhere, across any software program, platform, or notetaking tool. You can use the same system with the same categories and the same principles across your digital life.

You'll always need to use multiple platforms to move your projects forward. No single platform can do everything. The intention here is not to use a single software program, but to use a single *organizing system*, one that provides consistency even as you switch between apps many times per day. A project will be the same project whether it's found in your notes app, your computer file system, or your cloud storage drive, allowing you to move seamlessly between them without losing your train of thought.

By structuring your notes and files around the completion of your active projects, your knowledge can go to work for you, instead of collecting dust like an "idea graveyard." The promise of PARA is that it changes "getting organized" from a herculean, never-ending endeavor into a straightforward task to get over with so you can move on to more important work.

How PARA Works: Priming Your Mind (and Notes) for Action

With the PARA system, every piece of information you want to save can be placed into one of just four categories:

1. **Projects:** Short-term efforts in your work or life that you're working on now.

2. **Areas:** Long-term responsibilities you want to manage over time.

3. **Resources:** Topics or interests that may be useful in the future.

4. **Archives:** Inactive items from the other three categories.

Projects: What I'm Working on Right Now

Projects include the short-term outcomes you're actively working toward right now.

Projects have a couple of features that make them an ideal way to organize modern work. First, they have a beginning and an end; they take place during a specific period of time and then they finish. Second, they have a specific, clear outcome that needs to happen in order for them to be checked off as complete, such as "finalize," "green-light," "launch," or "publish."

A project-centric way of working comes naturally in the creative and performing arts. Artists have paintings, dancers have dances, musicians have songs, and poets have poems. These are clearly identifiable, discrete chunks of work. This project-centric approach is increasingly finding its way into all knowledge work, a trend named the "Hollywood model" after the way films are made.

As an article in the *New York Times*[3] explains, "A project is identified; a team is assembled; it works together for precisely as long as is needed to complete the task; then the team disbands… The Hollywood model is now used to build bridges, design apps or start restaurants." It is becoming more and more common for all of us to work across teams, departments, and even different companies to execute collaborative projects, and then once it's over, each go our own way.

Examples of projects could include:

- **Projects at work:** Complete web-page design; Create slide deck for conference; Develop project schedule; Plan recruitment drive.

- **Personal projects:** Finish Spanish language course; Plan vacation; Buy new living room furniture; Find local volunteer opportunity.

- **Side projects:** Publish blog post; Launch crowdfunding campaign; Research best podcast microphone; Complete online course.

If you are not already framing your work in terms of specific, concrete projects, making this shift will give you a powerful jump start to your productivity. Whether you're self-employed, at a large corporation, or somewhere in between, we are all moving toward a world of project-based work. Knowing which projects you're currently committed to is crucial to being able to prioritize your week, plan your progress, and say no to things that aren't important.

Areas: What I'm Committed to Over Time

As important as projects are, not everything is a project.

For example, the area of our lives called "Finances" doesn't have a definite end date. It's something that we will have to think about and manage, in one way or another, for as long as we live. It doesn't have a final objective. Even if you win the lottery, you'll still have finances to manage (and it will probably require a lot more attention!).

In our work lives, we have various ongoing areas we're responsible for, such as "product development," "quality control," or "human resources." These are the job responsibilities that we were hired to take on. Sometimes there are others that we officially or unofficially have taken ownership of over time.

Each of these is an example of an area of responsibility, and together they make up the second main category of PARA. All these areas, both personal and professional, require certain information to be handled effectively, but they're not the same as projects.

PROJECT	AREA
Lose 10 pounds	Health
Publish a book	Writing
Save 3 months of expenses	Finances
Create app mockup	Product design
Develop contract template	Legal

In the case of finances, you may have notes from calls with your financial advisor, receipts or invoices for business purchases, and your monthly household budget, among many other kinds of information. You might also have more speculative information to manage, like financial projections, research on personal finance software, and data on investment trends you're keeping an eye on.

For a work-related area like "product development," you might need to save product specifications, R&D findings, notes from customer research interviews, and customer satisfaction ratings. You could also have photos of products you admire to use as design inspiration, manufacturing blueprints, or color palettes. It all depends on your relationship to that area of your life, and how you want to manage it or move it forward.

Examples of areas from your personal life could include:

- **Activities or places you are responsible for:** Home/apartment; Cooking; Travel; Car.

- **People you are responsible for or accountable to:** Friends; Kids; Spouse; Pets.

- **Standards of performance you are responsible for:** Health; Personal growth; Friendships; Finances.

In your job or business:

- **Departments or functions you are responsible for:** Account management; Marketing; Operations; Product development.

- **People or teams you are responsible for or accountable to:** Direct reports; Manager; Board of directors; Suppliers.

- **Standards of performance you are responsible for:** Professional development; Sales and marketing; Relationships and networking; Recruiting and hiring.

Even though areas have no final outcome, it is still important to manage them. In fact, if you look at the list above, these areas are critical to your health, happiness, security, and life satisfaction.

While there is no goal to reach, there is a *standard* that you want to uphold in each of these areas. For finances, that standard may be that you always pay your bills on time and provide for your family's basic needs. For health, it may be that you exercise a certain number of times per week and keep your cholesterol below a certain number. For family, it may be that you spend quality time with them every evening and on the weekend.

Only you can decide what those standards are. For our purposes, it helps tremendously to have a place dedicated to each of them. That way you always have somewhere to put any thought, reflection, idea, or useful tidbit of information relevant to each important aspect of your life.

Resources: Things I Want to Reference in the Future

The third category of information that we want to keep is resources. This is basically a catchall for anything that doesn't belong to a project or an area and could include any topic you're interested in gathering information about.

For example:

- **What topics are you interested in?** Architecture; Interior design; English literature; Beer brewing.

- **What subjects are you researching?** Habit formation; Notetaking; Project management; Nutrition.

- **What useful information do you want to be able to reference?** Vacation itineraries; Life goals; Stock photos; Product testimonials.

- **Which hobbies or passions do you have?** Coffee; Classic movies; Hip-hop music; Japanese anime.

Any one of these subjects could become its own resource folder. You can also think of them as "research" or "reference materials." They are trends you are keeping track of, ideas related to your job or industry, hobbies and side interests, and things you're merely curious about. These folders are like the class notebooks you probably kept in school: one for biology, another for history, another for math. Any note or file that isn't relevant or actionable for a current project or area can be placed into resources for future reference.

Archives: Things I've Completed or Put on Hold

Finally, we have our archives. This includes any item from the previous three categories that is no longer active. For example:

- Projects that are completed or canceled

- Areas of responsibility that you are no longer committed to maintaining (such as when a relationship ends or after moving out of your apartment)

- Resources that are no longer relevant (hobbies you lose interest in or subjects you no longer care about)

The archives are an important part of PARA because they allow you to place a folder in "cold storage" so that it doesn't clutter your workspace, while safekeeping it forever just in case you need it. Unlike with your house or garage, there is no penalty for keeping digital stuff forever, as long as it doesn't distract from your day-to-day focus. If you need access to that information in the future—for example, if you take on a project similar to one you previously completed—you can call it up within seconds.

What PARA Looks Like: A Behind-the-Scenes Snapshot

PARA is a universal system of organization designed to work across your digital world. It doesn't work in only one place, requiring you to use completely different organizing schemes in each of the dozens of places you keep things. It can and should be used everywhere, such as the documents folder on your computer, your cloud storage drives, and of course, your digital notes app.

Let me show you what it looks like.

Here's an example of what the folders in my notes app look like with PARA:

Inside each of these top-level folders, I have individual folders for the specific projects, areas, resources, and archives that make up my life. For example, here are the folders for each one of my active projects:

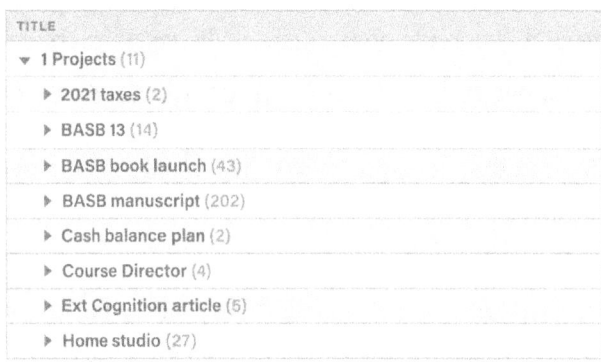

Inside these folders live the actual notes that contain my ideas. The number of active projects usually ranges from five to fifteen for the average person. Notice that the number of notes inside each one (indicated by the number in parentheses after the title) varies greatly, from just two to over two hundred for the book you're reading right now.

Here are the notes found within a typical project folder for a midsize project, a remodel of our garage into a home studio (which we'll dive deeper into in subsequent chapters):

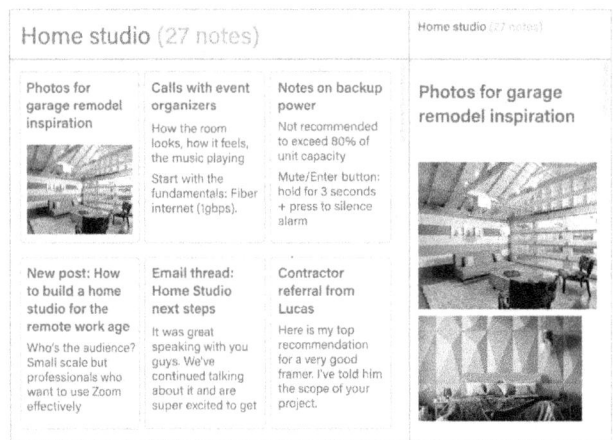

The left half of the window displays a list of the twenty-seven notes within this folder. Clicking a note, such as the one shown above containing a collection of photos we used to inspire our own remodel, reveals its contents in the right half of the window.

That's it—just three levels of hierarchy to encompass the thousands of notes I've accumulated over the years: the top-level PARA categories, the project folder, and the notes themselves.

Here is what some of my areas look like:

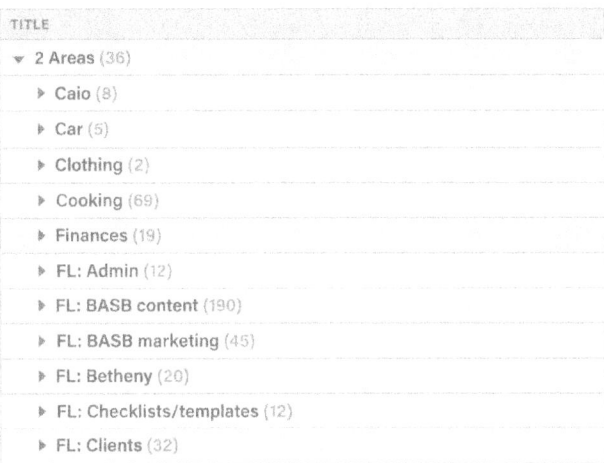

Each of these folders contains the notes relevant to each of those ongoing areas of my life. Areas related to my business begin with "FL" for Forte Labs, so they appear together in alphabetical order. Here are some of the notes in the "Health" area:

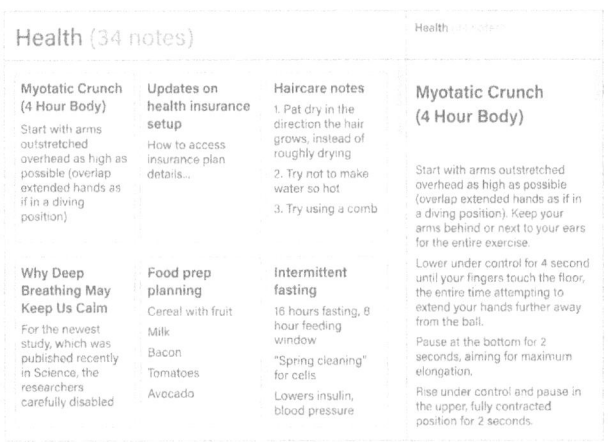

Under resources I have folders for each of the topics I'm interested in. This information isn't currently actionable, so I don't want it cluttering up my projects, but it will be ready and waiting if I ever need it.

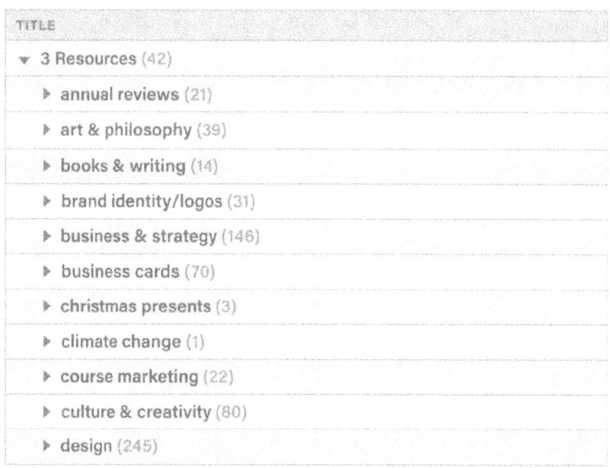

The archives contain any folder from the previous three categories that is no longer active. I want them completely out of sight and off my mind, but in case I ever need to access research, learnings, or material from the past, it will always be preserved.

PARA can be used across all the different places where you store information, meaning you can use the same categories and the same rules of thumb no matter where you keep content. For example, here is the documents folder on my computer:

And the folders for each one of my active projects:

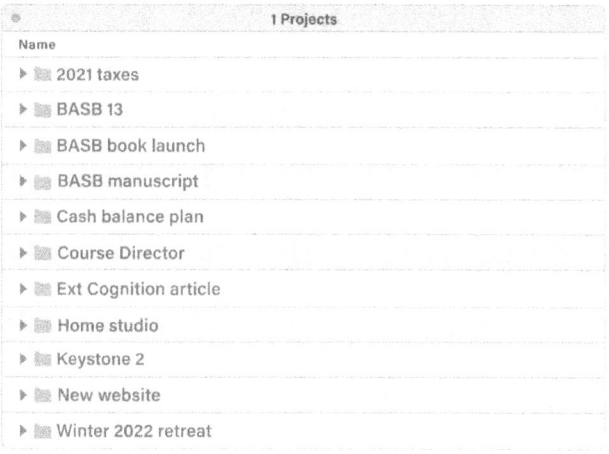

Inside these folders live the files that I use to execute each project. Here is the project folder dedicated to the book you're reading right now:

Where Do I Put This?—How to Decide Where to Save Individual Notes

Setting up folders is relatively easy. The harder question that strikes fear into the heart of every organizer is "Where do I put this?"

Apps have made it extremely easy to capture content—it's just a click or a tap away. However, we are given no guidance for what to do next. Where does a note go once it's been created? What is the correct location for an incoming file? The more material piles up, the more urgent and stressful this problem becomes.

The temptation when initially capturing notes is to also try to decide where they should go and what they mean. Here's the problem: the moment you first capture an idea is the worst time to try to decide what it relates to. First, because you've just encountered it and haven't had any time to ponder its ultimate purpose, but more importantly, because forcing yourself to make decisions every time you capture something adds a lot of friction to the process. This makes the experience mentally taxing and thus less likely to happen in the first place.

This is why it's so important to separate capture and organize into two distinct steps: "keeping what resonates" in the moment is a separate decision from deciding to save something for the long term. Most notes apps have an "inbox" or "daily notes" section where new notes you've captured are saved until you can revisit them and decide where they belong. Think of it as a waiting area where new ideas live until you are ready to digest them into your Second Brain. Separating the capturing and organizing of ideas helps you stay present, notice what resonates, and leave the decision of what to do with them to a separate time (such as a "weekly review," which I will cover in Chapter 9).

Once you've captured a batch of notes and it's time to organize them, PARA comes into play. The four main categories are ordered by actionability to make the decision of where to put notes as easy as possible:

- Projects are most actionable because you're working on them right now and with a concrete deadline in mind.

- Areas have a longer time horizon and are less immediately actionable.

- Resources may become actionable depending on the situation.

- Archives remain inactive unless they are needed.

This order gives us a convenient checklist for deciding where to put a note, starting at the top of the list and moving down:

1. In which project will this be most useful?
2. *If none*: In which area will this be most useful?
3. *If none*: Which resource does this belong to?
4. *If none*: Place in archives.

In other words, you are always trying to place a note or file not only where it will be useful, but where it will be useful the *soonest*. By placing a note in a project folder, you ensure you'll see it next time you work on that project. By placing it in an area folder, you'll come across it next time you're thinking about that area of your work or life. By placing it in a resource folder, you'll notice it only if and when you decide to dive into that topic and do some reading or research. By placing it in archives, you never need to see it again unless you want to.

It can be easy to let our projects and goals fall by the wayside when life gets busy. Personal projects and long-term goals feel especially flexible, like you can always get around to them later. Notes, bookmarks, highlights, and research that we worked hard to find sink deeper and deeper into our file systems, until eventually we forget they even exist.

Organizing by actionability counteracts our tendency to constantly procrastinate and postpone our aspirations to some far-off future. PARA pulls these distant dreams into the here and now, by helping us see that we already have a lot of the information we need to get started. The goal of organizing our knowledge is to move our goals forward, not get a PhD in notetaking. Knowledge is best applied through execution, which means whatever doesn't help you make progress on your projects is probably detracting from them.

Organizing Information Like a Kitchen—What Am I Making?

There is a parallel between PARA and how kitchens are organized.

Everything in a kitchen is designed and organized to support an outcome—preparing a meal as efficiently as possible. The archives are like the freezer—items are in cold storage until they are needed, which could be far into the future. Resources are like the pantry—available for use in any meal you make, but neatly tucked away out of sight in the meantime. Areas are like the fridge—items that you plan on using relatively soon, and that you want to check on more frequently. Projects are like the pots and pans cooking on the stove—the items you are actively preparing right now. Each kind of food is organized according to how accessible it needs to be for you to make the meals you want to eat.

Imagine how absurd it would be to organize a kitchen instead by *kind of food*: fresh fruit, dried fruit, fruit juice, and frozen fruit would all be stored in the same place, just because they all happen to be made of fruit. Yet this is exactly the way most people organize their files and notes—keeping all their book notes together just because they happen to come from books, or all their saved quotes together just because they happen to be quotes.

Instead of organizing ideas according to *where they come from*, I recommend organizing them according to *where they are going*—specifically, the outcomes that they can help you realize. The true test of whether a piece of knowledge is valuable is not whether it is perfectly organized and neatly labeled, but whether it can have an impact on someone or something that matters to you.

PARA isn't a filing system; it's a production system. It's no use trying to find the "perfect place" where a note or file belongs. There isn't one. The whole system is constantly shifting and changing in sync with your constantly changing life.

This is a challenging idea for a lot of people to wrap their head around. We are used to organizational systems that are static and fixed. We expect to find a strict set of rules that tells us exactly where each item goes, like the precise call numbers for books in a library.

When it comes to our personal knowledge, there is no such assigned spot. We are organizing for actionability, and "what's actionable" is always changing. Sometimes we can receive one text message or email and the entire landscape of our day changes. Because our priorities can change at a moment's notice, we have to minimize the time we spend filing, labeling, tagging, and maintaining our digital notes. We can't run the risk of all that effort going to waste.

Any piece of information (whether a text document, an image, a note, or an entire folder) can and should flow between categories. You might save a note on coaching techniques to a project folder called "Coaching class," for a class you're taking. Later, when you become a manager at work and need to coach your direct reports, you might move that note to an area folder called "Direct reports." At some point you might leave that company, but still remain interested in coaching, and move the note to resources. One day you might lose interest in the subject altogether and move it to the archives. In the future, that note could find its way all the way back to projects when you decide to start a side gig as a business coach, making that knowledge actionable once again.

The purpose of a single note or group of notes can and does change over time as your needs and goals change. Every life moves through seasons, and your digital notes should move along with them, churning and surfacing new tidbits of insight from the deep waters of your experience.

Completed Projects Are the Oxygen of Your Second Brain

Your efforts to capture content for future use will be tremendously easier and more effective if you know what that content is for. Using PARA is not just about creating a bunch of folders to put things in. It is about identifying the structure of your work and life—what you are committed to, what you want to change, and where you want to go.

I had to learn this lesson the hard way. Back in college, I worked part-time at an Apple Store in San Diego while I finished my studies. At the time it was one of the five busiest Apple Stores in the world, with thousands of people walking through our doors every day. It was there that I got my first taste of teaching people how to use computers more effectively.

I taught morning classes to small groups of people who had just bought their first Mac, and also did one-on-one consultation sessions. This was the golden age of Apple's iLife suite of creative software: every single Mac computer came preinstalled with user-friendly apps for creating websites, recording music, printing photo books, and making videos. It was like having a complete multimedia studio at your fingertips at no additional cost.

I would sit down with customers and answer any questions they had about the new computer they had just bought. In most cases they had just migrated all their files over from Windows, and years of accumulated documents lay scattered across their desktop and documents folders.

At first I tried guiding them through organizing each document one at a time. It quickly became clear that this didn't work at all. The one-on-one sessions were only one hour, not nearly enough time to make even a dent in the hundreds or even thousands of files they had. It wasn't time well spent anyway, because these were often old documents that weren't relevant to their current goals or interests.

I knew I needed a new approach. I started asking questions and listening, and eventually realized that these people didn't need or want an organized computer. They had spent all this money and time moving to a Mac because there was something they wanted to create or achieve.

They wanted to make a video for their parents' anniversary party, a website for their cupcake shop, or a record showcasing their band's songs. They wanted to research their family genealogy, graduate from college, or land a better job. Everything else was just an obstacle to get past on the way to their goal.

I decided to take a different approach: I took all the files they'd migrated over and moved them all to a new folder titled "Archive" plus the date (for example, "Archive 5-2-21"). There was always a moment of fear and hesitation at first. They didn't want anything to get lost, but very quickly, as they saw that they would always be able to access anything from the past, I watched them come alive with a renewed sense of hope and possibility.

They had repeatedly postponed their creative ambitions to some far-off, mythical time when somehow everything would be perfectly in order. Once we set that aside and just focused on what they actually wanted to do right now, they suddenly gained a tremendous sense of clarity and motivation.

For a while I was sure this would come back to haunt me. Eventually they'd want to go back and organize all those old files, right? I would often see the same people coming back to our store again and again. I waited expectantly in fear for someone to return and accuse me of losing all their old files.

Let me tell you: no one ever did.

Not once did someone come back and say, "You know, I'd really like to go back and organize all those files from my old computer." What they did tell me were the stories of the impact their creative projects had: on their families, on their business, on their grades, on their career. One person organized a fundraising drive for a friend who had recently been diagnosed with leukemia. Another put together a successful application for a small business loan to start a dance studio. One student told me that the ability to tame the chaos of her digital world was the only reason she had finished college as the first graduate from her family. The details of how they organized their computer or took notes were trivial, but the impact their creativity had on their own lives and the lives of others—that was anything but.

There are a few lessons I took away from this experience.

The first is that people need clear workspaces to be able to create. We cannot do our best thinking and our best work when all the "stuff" from the past is crowding and cluttering our space. That's why that archiving step is so crucial: you're not losing anything, and it can all be found via search, but you need to move it all out of sight and out of mind.

Second, I learned that creating new things is what really matters. I'd see a fire light up in people's eyes when they reached the finish line and published that slideshow or exported that video or printed that résumé. The newfound confidence they had in themselves was unmistakable as they walked out of the store knowing they had everything they needed to move forward.

I've learned that completed creative projects are the blood flow of your Second Brain. They keep the whole system nourished, fresh, and primed for action. It doesn't matter how organized, aesthetically pleasing, or impressive your notetaking system is. It is only the steady completion of tangible wins that can infuse you with a sense of determination, momentum, and accomplishment. It doesn't matter how small the victories. Even the tiniest breakthrough can become a stepping-stone to more creative, more interesting futures than you can imagine.

Your Turn: Move Quickly, Touch Lightly

A mentor of mine once gave me a piece of advice that has served me ever since: move quickly and touch lightly.

She saw that my standard approach to my work was brute force: to stay late at the office, fill every single minute with productivity, and power through mountains of work as if my life depended on it. That wasn't a path to success; it was a path to burnout. Not only did I exhaust my mental and physical reserves time and again; my frontal assaults weren't even very effective. I didn't know how to set my intentions, craft a strategy, and look for sources of leverage that would allow me to accomplish things with minimal effort.

My mentor advised me to "move quickly and touch lightly" instead. To look for the path of least resistance and make progress in short steps. I want to give the same advice to you: don't make organizing your Second Brain into yet another heavy obligation. Ask yourself: "What is the smallest, easiest step I can take that moves me in the right direction?"

When it comes to PARA, that step is generally to create folders for each of your active projects in your notes app and begin to fill them with the content related to those projects. Once you have a home for something, you tend to find more of it. Start by asking yourself, "What projects am I currently committed to moving forward?" and then create a new project folder for each one. Here are some questions to ask yourself to help you think of the projects that might be on your plate:

- **Notice what's on your mind:** What's worrying you that you haven't taken the time to identify as a project? What needs to happen that you're not making consistent progress on?

- **Look at your calendar:** What do you need to follow up on from the past? What needs planning and preparation for the future?

- **Look at your to-do list:** What actions are you already taking that are actually part of a bigger project you've not yet identified? What communication or follow-up actions you've scheduled with people are actually part of a bigger project?

- **Look at your computer desktop, downloads folder, documents folder, bookmarks, emails, or open browser tabs:** What are you keeping around because it is part of a larger project?

Here's some examples of projects my students have come up with:

- Find new doctor who accepts my insurance
- Plan goals and agenda for annual team retreat
- Collect list of common food supplies and set up recurring deliveries
- Develop content strategy for next quarter
- Review draft of new reimbursement policy and provide feedback
- Share collaboration ideas with research partner
- Research and draft article on health equity
- Complete online course on creative writing

You could also create folders for your areas and resources, but I recommend starting only with projects to avoid creating lots of empty containers. You can always add others later when you have something to put inside them. Although you can and should use PARA across all the platforms where you store information—the three most common ones besides a notetaking app are the documents folder on your computer, cloud storage drives like Dropbox, and online collaboration suites like Google Docs—I recommend starting with just your notes app for now.

Practice capturing new notes, organizing them into folders, and moving them from one folder to another. Each time you finish a project, move its folder wholesale to the archives, and each time you start a new project, look through your archives to see if any past project might have assets you can reuse.

As you create these folders and move notes into them, don't worry about reorganizing or "cleaning up" any existing notes. You can't afford to spend a lot of time on old content that you're not sure you're ever going to need. Start with a clean slate by putting your existing notes in the archives for safekeeping. If you ever need them, they'll show up in searches and remain just as you left them.

Your goal is to clear your virtual workspace and gather all the items related to each active project in one place. Once you do, you'll gain the confidence and clarity to take action on those ideas, rather than letting them pile up with no end in sight.

The key thing to keep in mind is that these categories are anything but final. PARA is a dynamic, constantly changing system, not a static one. Your Second Brain evolves as constantly as your projects and goals change, which means you never have to worry about getting it perfect, or having it finished.

In the next chapter, we'll look at how to distill the knowledge we've gathered to be able to put it to use effectively.

I. *Para* is a Greek word that means "side by side," as in "parallel"; this convenient definition reminds us that our Second Brain works "side by side" with our biological brain.

II. As you can probably tell, I'm a big fan of four-letter frameworks. Researchers have called it the "Magic Number 4" because it is the highest number that we can count at a glance and hold in our minds without extra effort.

III. I will use the term "folder" to refer to the main unit of organization used by most notes apps; some software instead uses tags, which work just as well.

Chapter 6

Distill—Find the Essence

> To attain knowledge, add things every day. To attain wisdom, remove things every day.
> —Lao Tzu, ancient Chinese philosopher

In 1969, studio executives at Paramount Pictures were desperate to find a film director for a new movie they had purchased the rights to, a crime drama based on the New York Mafia.

One after another, all the top directors of the era turned the project down. They all found it too sensationalistic for their tastes. Gangster movies were known as cliché and gimmicky, and there had been several recent duds in the genre.

After exhausting all their top choices, studio executives approached a young film director who had done a few small indie films. The director was a relative novice, with no major commercially successful films to his name. He was an outsider, working out of San Francisco instead of Hollywood, the industry capital. And he was known as an artist who wanted to experiment with new ideas, not a director of big budget movies.

That director's name was Francis Ford Coppola, and the movie he was asked to make was, of course, *The Godfather*.

Coppola initially turned down the project. As recounted in the *Hollywood Reporter*,[1] he said, "It was more commercial and salacious than my own taste." However, his partner and protégé George Lucas (of future *Star Wars* fame) noted that they were broke. Without a major infusion of cash, they'd soon be evicted.

Mounting financial pressure, plus a second reading of the novel, changed Coppola's mind, as he realized that it could be framed as "a story that was a metaphor for American capitalism in the tale of a great king with three sons."

The Godfather would go on to become one of the greatest critical and commercial successes in filmmaking history. In 2007 the American Film Institute named it the third-best American movie of all time.[2] It ultimately grossed $245 million, won three Oscars, and spawned a series of sequels and spinoffs eaten up by a rabid fan base obsessed with the story of the fictional Corleone family.

Coppola's strategy for making the complex, multifaceted film rested on a technique he learned studying theater at Hofstra College, known as a "prompt book." He started by reading *The Godfather* novel and capturing the parts that resonated with him in a notebook—his own version of Twyla Tharp's box. But his prompt book went beyond storage: it was the starting point for a process of revisiting and refining his sources to turn them into something new.

The book was made from a three-ring binder, into which he would cut and paste pages from the novel on which the film was based. It was designed to last, with reinforced grommets to ensure the pages wouldn't tear even after many turnings. There he could add the notes and directions that would later be used to plan the screenplay and production design of the film.

In a short documentary titled *Francis Coppola's Notebook*[3] released in 2001, Coppola explained his process. He started with an initial read of the entire novel, noting down anything that stuck out to him: "I think it's important to put your impressions down on the first reading because those are the initial instincts about what you thought was good or what you didn't understand or what you thought was bad."

Coppola then began to add his own interpretations, distilling and reconstituting his own version of the story. He broke down each scene according to five key criteria: a synopsis (or summary) of the scene; the historical context; the imagery and tone for the "look and feel" of a scene; the core intention; and any potential pitfalls to avoid. In his own words, "I endeavored to distill the essence of each scene into a sentence, expressing in a few words what the point of the scene was."

Coppola described his binder as "a kind of multi-layered road map for me to direct the film… so I was able to review not only Mario Puzo's original text but all my first notation as to what… was important to me or what I felt was really going on in the book." His comments in the margins included "Hitchcock" to remind him of how the famed director of thrillers would have framed a shot, or "Frozen time" to remind him to slow down a sequence. He used different kinds of annotations to emphasize to his future self which parts of a scene were most important: "As I was reading the book and making these notes and then putting them on the margins obviously the more pens I was using and the more rulers, and the more squiggly lines, sort of implied the excitement of the book was higher and higher, so that the sheer amount of ink on the page would tell me later on this is one of the most important scenes."

The Godfather Notebook is a perfect example of the behind-the-scenes process used by successful creative professionals. Coppola considered the prompt book that emerged from this process the most important asset in the production of his now classic film: "the script was really an unnecessary document; I didn't need a script because I could have made the movie just from this notebook."

We might imagine a movie as emerging straight out of the mind of a screenplay writer or director, when in fact it depends on collecting and refining source material. Coppola's story demonstrates that we can systematically gather building blocks from our reading and research that ultimately make the final product richer, more interesting, and more impactful.

If Francis Ford Coppola relied so much on a step-by-step process for notetaking, then so can we. We can also use our notes to drill down to the essence of the stories, research, examples, and metaphors that make up our own source material. This is the third step of CODE, to Distill. This is the moment we begin turning the ideas we've captured and organized into our own message. It all begins and ends with notes.

Quantum Notetaking: How to Create Notes for an Unknown Future

I've shown you how to capture notes with interesting ideas from the outside world or your own thoughts. You may have started organizing those notes according to their actionability and relevance to your current projects.

Now what?

This is where even the most dedicated notetakers usually stop. They aren't sure what to do next. They've gathered some interesting knowledge, but it hasn't led anywhere. Our notes are things to use, not just things to collect.

When you initially capture a note, you may have only seconds to get it into your Second Brain before the next meeting, urgent task, or crying child comes calling. Not nearly enough time to fully understand what it means or how it might be used. When you first capture them, your notes are like unfinished pieces of raw material. They require a bit more refinement to turn them into truly valuable knowledge assets, like a chemist distilling only the purest compound. This is why we separate capturing and organizing from the subsequent steps: you need to be able to store something quickly and save any future refinement for later.

In this sense, notetaking is like time travel—you are sending packets of knowledge through time to your future self.

You probably consume a lot of books, articles, videos, and social media posts full of interesting insights, but what are the chances that you'll be ready to put any given piece of advice into action right at that instant? How likely is it that life will intervene, in the form of a crisis at work, an urgent meeting at your kid's school, or an unexpected cold? In my experience, life is constantly pushing and pulling us away from our priorities. The more determined we are to focus and get something done, the more aggressively life tends to throw emergencies and delays in our face.

You're watching a YouTube video on home renovation now, but that knowledge can be put to use only in a few months, when you move into your new place. You're reading an article about time management techniques now, but they end up being most useful at the end of the year, when your new child is born and you suddenly have much greater demands on your time. You're talking to a sales prospect about their goals and challenges now, but when you could really use that information is next year, when they start taking bids for a huge new contract.

This holds true for so much of the ideas and inspiration around us. There is a key idea that catches our attention in the moment. We feel enraptured and obsessed with it. It's difficult to imagine ever forgetting the new idea. It's changed our lives forever! But after a few hours or days or weeks, it starts to fade from our memory. Soon our recollection of that exciting new idea is nothing but a pale shadow of something we once knew, that once intrigued us. Your job as a notetaker is to preserve the notes you're taking on the things you discover in such a way that they can survive the journey into the future. That way your excitement and enthusiasm for your knowledge builds over time instead of fading away.

Discoverability—The Missing Link in Making Notes Useful

The most important factor in whether your notes can survive that journey into the future is their *discoverability*—how easy it is to discover what they contain and access the specific points that are most immediately useful.

Discoverability is an idea from information science that refers to "the degree to which a piece of content or information can be found in a search of a file, database, or other information system."[1] Librarians think about discoverability when deciding how to lay out books on the shelves. Web designers think about it when they create menus for the websites you visit every day. Social media platforms work hard to make the best content on their platforms as discoverable as possible.

Discoverability is the element most often missing from people's notes. It's easy to save tons and tons of content, but turning it into a form that will be accessible in the future is another matter. To enhance the discoverability of your notes, we can turn to a simple habit you probably remember from school: highlighting the most important points. Highlighting is an activity that everyone understands, takes hardly any additional effort, and works in any app you might use.

Imagine your future self as a demanding customer. They will surely be impatient and very busy. They won't have time to pore through page after page of details just to find the hidden gems. It's your job to "sell" them on the value of the notes you are taking now. Your future self might have mere minutes before a meeting starts to quickly search their notes for a reference they need. In that sense, each note is like a product you are creating for the benefit of that future customer. If they don't buy it—they don't think it's worth the effort of revisiting past notes—then all the value of the work you're doing now will be lost.

This points to a paradox that a lot of people experience as they take notes: the more notes they gather, the more the volume of information grows, the more time and effort it takes to review it all, and the less time they have to do so. Paradoxically, the more notes they collect, the less discoverable they become! This realization tends to either discourage them from taking any notes in the first place, or alternatively, to keep switching from one notetaking tool to another every time the volume gets overwhelming. Thus, they miss out on most of the benefits of their knowledge compounding over time.

What do you do when communicating with a very busy, very impatient, very important person? You distill your message down to the key points and action steps. When you send an email to your boss, you don't bury your request somewhere near the bottom of a massive wall of text. You identify the most urgent questions that you need them to respond to right at the top of the email. When you're giving a presentation to the leadership of your organization, you don't drone on for hours. You leave out unnecessary details and get right to the point.

Distillation is at the very heart of all effective communication. The more important it is that your audience hear and take action on your message, the more distilled that message needs to be. The details and subtleties can come later once you have your audience's attention.

What if your future self was just as important as these VIPs? How could you communicate with them through time in the most efficient, concise way?

Highlighting 2.0: The Progressive Summarization Technique

Progressive Summarization is the technique I teach to distill notes down to their most important points. It is a simple process of taking the raw notes you've captured and organized and distilling them into usable material that can directly inform a current project.

Progressive Summarization takes advantage of a tool and a habit that we are all intimately familiar with—highlighting—while leveraging the unique capabilities of technology to make those highlights far more useful than anything you did in school.

The technique is simple: you highlight the main points of a note, and then highlight the main points of *those* highlights, and so on, distilling the essence of a note in several "layers." Each of these layers uses a different kind of formatting so you can easily tell them apart.

Here is a snapshot of the four layers of Progressive Summarization:[II]

Here's an example of a note I captured from an article in *Psychology Today*.[4] I came across the link when it was shared on social media and saved it with two clicks to my read later app, where I collect bookmarks of everything I want to read, watch, or listen to. A few evenings later, when I wanted to do some casual reading to wind down for the day, I read the article and highlighted the passages I found most interesting. I have my read later app synced to my digital notes app, so any passage I highlight there automatically gets saved in my notes, including a link to the source.

> **How the Brain Stops Time**
>
> One of the strangest side effects of intense fear is time dilation, the apparent slowing down of time...survivors of life-and-death situations often report that things seem to take longer to happen, objects fall more slowly, and they're capable of complex thoughts in what would normally be the blink of an eye.
>
> Eagleman asked subjects who'd already taken the plunge to estimate how long it took them to fall, using a stopwatch to tick off what they felt to be an equivalent amount of time. Then he asked them to watch someone else fall and then estimate the elapsed time for their plunge in the same way. On average, participants felt that their own experience had taken 36 percent longer. Time dilation was in effect.
>
> That means that fear does not actually speed up our rate of perception or mental processing. Instead, it allows us to remember what we do experience in greater detail. Since our perception of time is based on the number of things we remember, fearful experiences thus seem to unfold more slowly.
>
> Source link

This is what I call "layer one"—the chunks of text initially captured in my notes. Notice that I didn't save the entire article—only a few key excerpts.[III] By limiting what I keep to only the best, most important, most relevant parts, I'm making all the subsequent steps of organizing, distilling, and expressing much easier. If I ever need to know the full details, I have the link to the original article right there at the bottom.

As interesting as this content is, it's not nearly succinct enough. Once again, in the midst of a chaotic workday, I would be hard-pressed to find the time to casually look through multiple paragraphs of text to find the relevant points. Unless I highlight those points in a way that my future self can instantly grasp, I'll likely never see them again.

To enhance the discoverability of this note, I need to add a second layer of distillation. I usually do this when I have free time during breaks or on evenings or weekends, when I come across the note while working on other projects, or when I don't have the energy for more focused work. All I have to do is bold the main points within the note. This could include keywords that provide hints of what this text is about, phrases that capture what the original author was trying to say, or sentences that especially resonated with me even if I can't explain why. Looking over the bolded parts of the same note below, can you see how much easier it is to quickly grasp the gist of this note by looking only at those parts?

At layer two, this note is already dramatically more discoverable. Imagine the difference between reading the original article, which might take five to ten minutes of focused attention, versus glancing over these bolded points, which would take less than a minute.

> **How the Brain Stops Time**
>
> **One of the strangest side effects of intense fear is time dilation, the apparent slowing down of time**…survivors of life-and-death situations often report that things seem to take longer to happen, objects fall more slowly, and they're **capable of complex thoughts in what would normally be the blink of an eye.**
>
> Eagleman **asked subjects who'd already taken the plunge to estimate how long it took them to fall**, using a stopwatch to tick off what they felt to be an equivalent amount of time. Then he asked them to watch someone else fall and then estimate the elapsed time for their plunge in the same way. **On average, participants felt that their own experience had taken 36 percent longer. Time dilation was in effect.**
>
> That means that fear does not actually speed up our rate of perception or mental processing. **Instead, it allows us to remember what we do experience in greater detail.** Since our perception of time is based on the number of things we remember, **fearful experiences thus seem to unfold more slowly.**
>
> Source link

We're not done yet! For those notes that are especially long, interesting, or valuable, it is sometimes worth adding a third layer of highlighting. I advise using the "highlighting" feature offered by most notes apps, which paints passages in bright yellow just like the fluorescent highlighters we used in school (which appear in light gray below). If your notes app doesn't have a highlighting feature, you can use underlining or another kind of formatting instead. Look only at the bolded passages you identified in layer two and highlight only the most interesting and surprising of those points. This will often amount to just one or two sentences that encapsulate the message of the original source.

> **How the Brain Stops Time**
>
> **One of the strangest side effects of intense fear is time dilation, the apparent slowing down of time**...survivors of life-and-death situations often report that things seem to take longer to happen, objects fall more slowly, and they're **capable of complex thoughts in what would normally be the blink of an eye.**
>
> Eagleman **asked subjects who'd already taken the plunge to estimate how long it took them to fall,** using a stopwatch to tick off what they felt to be an equivalent amount of time. Then he asked them to watch someone else fall and then estimate the elapsed time for their plunge in the same way. **On average, participants felt that their own experience had taken 36 percent longer. Time dilation was in effect.**
>
> That means that fear does not actually speed up our rate of perception or mental processing. **Instead, it allows us to remember what we do experience in greater detail.** Since our perception of time is based on the number of things we remember, **fearful experiences thus seem to unfold more slowly.**
>
> Source link

Looking at the note above, can you see how those few highlighted sentences jump out and catch your eye? They convey the main message of this article in a highly distilled form that takes just seconds to grasp. When I come across this note in the future—while doing a search or browsing the notes within a folder—I'll be able to decide in the blink of an eye whether this source is relevant to my needs. If it is, I'll have all the additional details and context I need to remember it right in front of me, as well as the link to the original article to check the source.

There is one more layer we can add, though it is quite rarely needed. For only the very few sources that are truly unique and valuable, I'll add an "executive summary" at the top of the note with a few bullet points summarizing the article in my own words. The best sign that a fourth layer is needed is when I find myself visiting a note again and again, clearly indicating that it is one of the cornerstones of my thinking. Looking only at the points I've previously bolded and highlighted in layers two and three makes it far easier to write this summary than if I was trying to summarize the entire article all at once.

I recommend using bullet points to encourage yourself to make this executive summary succinct. Use your own words, define any unusual terms you're using, and think about how your future self, who may not remember anything about this source, might interpret what you're writing.

> **How the Brain Stops Time**
>
> **Summary**
> - Time dilation is the feeling that time is slowing down
> - It is often experienced during moments of intense fear
> - In an experiment, subjects experienced time moving 36% slower in a state of fear, compared to watching others' experience
> - Further experiments showed that time dilation allows us to remember our experiences better
>
> **How the Brain Stops Time**
>
> **One of the strangest side effects of intense fear is time dilation, the apparent slowing down of time**...survivors of life-and-death situations often report that things seem to take longer to happen, objects fall more slowly, and they're capable of complex thoughts in what would normally be the blink of an eye.
>
> Eagleman asked subjects who'd already taken the plunge to estimate how long it took them to fall, using a stopwatch to tick off what they felt to be an equivalent amount of time. Then he asked them to watch someone else fall and then estimate the elapsed time for their plunge in the same way. **On average, participants felt that their own experience had taken 36 percent longer. Time dilation was in effect.**
>
> That means that fear does not actually speed up our rate of perception or mental processing. **Instead, it allows us to remember what we do experience in greater detail.** Since our perception of time is based on the number of things we remember, fearful experiences thus seem to unfold more slowly.
>
> Source link

By reviewing this executive summary, I can rapidly recall the main takeaways from this article in a fraction of the time it would take to reread the original. Since the takeaways are already in my own words, they are easy to incorporate into whatever I'm working on. Speed is everything when it comes to recall: you have only a limited amount of time and energy, and the faster you can move through your notes, the more diverse and interesting ideas you can connect together.

Zooming In and Out of Your Map of Knowledge

The layers of Progressive Summarization give you multiple ways of interacting with your notes depending on the needs of the moment. The first time you read about a new idea, you might want to dive into the details and explore every nuance. The next time you revisit that idea, you probably don't want to repeat all that effort and read the same piece from beginning to end again. You want to pick up where you left off, looking only at the highlights left over from the last time you visited that note. You can review all the details at layer one, or if you're pressed for time (and when are we ever not pressed for time?), just focus on layers two, three, or four. You can *customize* how much attention you spend on a note based on your energy level and time available.

It's like having a digital map of your notes that can be zoomed in or out depending on how many details you want to see, like a maps app on your smartphone. Navigating to a new destination, you might want to zoom in and see exactly which driveway to turn into. On the other hand, if you're planning a cross-country road trip, you might want to zoom out and see your entire itinerary in one glance. The same is true for your landscape of knowledge—sometimes you want to zoom in and examine one specific research finding, while other times you want to zoom out and see the broad sweep of an argument all at once.

With Progressive Summarization, you are building up a map of the best ideas found in your Second Brain. Your highlights are like signposts and waypoints that help you navigate through the network of ideas you're exploring. You are building this map without moving anything or deleting anything. Every sentence gets left right where you found it, giving you the freedom to leave things out without worrying that you'll lose them. With this map in hand, you can actually see what you've captured, helping you find what you're looking for but also what you don't even know you're looking for.

Highlighting can sometimes feel risky. You may wonder, "Am I making the right decision about which points are most important, or what this source means?" The multiple layers of Progressive Summarization are like a safety net; if you go in the wrong direction, or make a mistake, you can always just go back to the original version and try again. Nothing gets forgotten or deleted.

Progressive Summarization helps you focus on the *content* and the *presentation* of your notes,[IV] instead of spending too much time on labeling, tagging, linking, or other advanced features offered by many information management tools. It gives you a practical, easy thing to do that adds value even when you don't have the energy for more challenging tasks. Most importantly, it keeps your attention on the substance of what you're reading or learning, which is what matters in the long term.

Four Examples of Progressive Summarization

Progressive Summarization can be used across a wide variety of different kinds of content. As long as a source can be turned into text,[v] you can add layers of highlighting in any information management tool you use.

Let's look at more examples of progressively summarized notes:

- A Wikipedia article
- A blog post
- A podcast interview
- Meeting notes

Wikipedia Articles

Have you ever found yourself visiting the same Wikipedia article again and again or trying to remember something from that one article you read weeks ago?

By saving the best excerpts from Wikipedia articles you read, you can create your own private encyclopedia with only the parts that are most relevant to you. In the note below, I captured a few key sentences from the article explaining "Baumol's Cost Disease," a somewhat esoteric term from economics that I'd seen referenced a few times.

When I first captured the note, I didn't have time to add tags, highlights, or an executive summary of my own. I saved it to a resource folder for "Economics" to revisit later. A few months later, when it came up in a search for "wages," I took a few moments to bold a couple of key sentences and highlight the most important one, so I could get the gist of it with a glance.

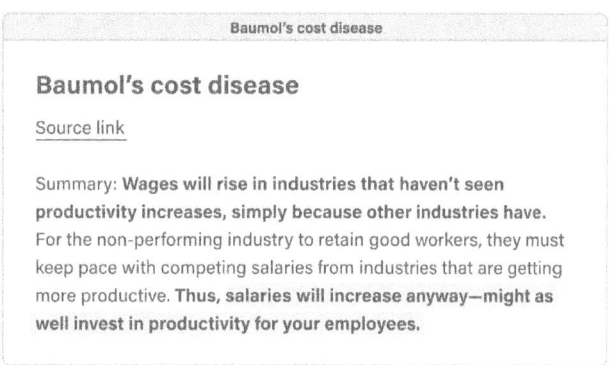

I was once on a panel when one of the speakers mentioned this term. Within the ten seconds before it was my turn to respond, I was able to run a search, look up this note on my tablet (where all my notes are synchronized), and speak confidently on the subject as if I had known it all along.

Online Articles

Much of the time we consume information without a specific purpose in mind. We might peruse the newspaper over breakfast, listen to a podcast while working out, or check out a newsletter to casually learn about a topic. We consume information to stay up to date, pass the time, entertain ourselves, and keep our minds engaged.

These moments are some of the most valuable opportunities to capture tidbits of insight that you probably might not find otherwise. Because this casual reading and listening tends to range over a wide number of topics and interests, you are exposed to more diverse ideas than usual.

One evening I was reading an online article that I saw shared on social media. The article explained how Google used "structured interviews" as part of their hiring process to reduce bias, ensure consistency, and learn from past hires. I was a solo freelancer at the time and had no immediate use for knowledge about hiring practices. I knew that someday I might, so I decided to save the paragraph you see below to my Second Brain.

> **Hiring Process and Retention Rates | FLOX**
>
> Source link
>
> "Fortunately, research also shows that **structured interviews—simply using the same interview questions and techniques to evaluate job candidates—drastically reduces our biases when hiring.** Compared to unstructured interviews, structured interviews are also: **Better for diversity.** You get smaller differences in interview scores between different ethnicities. More efficient. Your questions and criteria are set for the 100+ candidates you interview. More liked by job candidates. Applicants who like your hiring process significantly predict higher job performance by about 10%."

Almost two years later, I was finally ready to hire my first employee. I remember the feeling of anxiety as I prepared to take on this major financial commitment, not to mention the responsibility of managing a direct report. Luckily, I had a handful of highly actionable notes saved in a resource folder called "Hiring." To get started, I moved the entire folder from resources to projects. Then I spent about thirty minutes to review the notes it contained and highlight the most relevant takeaways. Those highlights were the starting point for the hiring process I ended up using for my own business, inspired by one of the most innovative and desirable employers in the world.

Podcasts and Audio

Notes can come in handy even when you're not able to write them down in real time. I was driving with my wife one weekend to a small Airbnb cabin in the Sierra Nevada mountains of California, and we decided to listen to a podcast. It was a casual conversation between the host and a course instructor named Meghan Telpner, who ran an online school called the Academy of Culinary Nutrition.[5]

I had never heard of her and put on the episode without any particular goal in mind. Over the next hour, as we ascended the steep mountainside roads, we were captivated by the story of the education business she had managed to build. She had faced so many of the same challenges we had. It was a relief to hear that we weren't alone in our struggles. I was driving and unable to write anything down, but as soon as we arrived I sat in the car for a few minutes and captured the ideas I remembered. This is actually a great way to filter down the volume of notes you're taking—the best stuff always sticks in your mind for an hour or two.

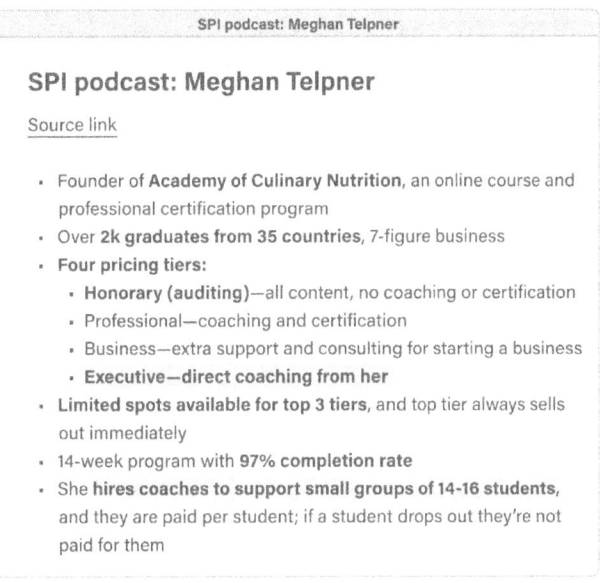

A few months later, we were preparing a launch campaign for a new version of our online course. I had only a couple of weeks to prepare for it—definitely not enough time to do more research. I had to make use of the ideas I'd already collected. As part of my preparation, I went through this note (which I found within an area folder for "Online education") and bolded the parts that most resonated with me. Then just before our launch kickoff I highlighted the parts I wanted to apply to our own situation. The highlighted passages you see here were the sparks that eventually led to us hiring alumni of the course to coach new students. This freed up my time to implement another idea from Telpner's interview: adding a new "executive" coaching tier. You truly never know where inspiration will come from and the extraordinary impact it can have.

Meeting Notes

Like many people, I spend a fair percentage of my time on phone calls and in meetings. I want to make the best use of that time, so I take notes during most meetings of new ideas, suggestions, feedback, and action steps that come up.

Taking notes during meetings is a common practice, but it's often not clear what we should do with those notes. They are often messy, with the action items buried among random comments. I often use Progressive Summarization to summarize my notes after phone calls to make sure I'm extracting every bit of value from them.

I captured this note during a conversation with a friend of mine who has experience designing recording studios. We were remodeling our garage into a home studio and I wanted to get his advice. He was kind enough to come over and walk me through his recommendations, and I wrote down the main points in a notes app on my smartphone as he spoke.

Derick's home studio recommendations

- 4 section bifold door with frosted glass
- Black theater blackout curtain that can be pulled over the entire inside door (to block both light and echoes); eyelets across the top so we can hang it or take it off when not in use; OR have a baffle at the corner of the garage to store the curtain and block edges of light
- Modular carpet squares with cable tracks underneath them
- Opening ceiling completely and painting it all black; put in pipe trussing so we can hang lights, cameras from the ceiling; OR cable ceiling tracks so we don't have to worry about zip ties
- Sound absorption panels in black placed around ceiling so they disappear; hung with wood screws and washers

Sometime later, I happened to be driving by the local hardware store on my way home. I realized that I could pop in and get some of the supplies my friend had recommended. I took out my smartphone, did a search for "home studio," and found this note. I took a few minutes while sitting in the car and bolded the items I would need to purchase at some point, which were buried among other suggestions he had made.

Here's what it looked like:

> **Derick's home studio recommendations**
>
> - **4 section bifold door** with frosted glass
> - Black theater **blackout curtain** that can be pulled over the entire inside door (to block both light and echoes); **eyelets across the top** so we can hang it or take it off when not in use; OR have a baffle at the corner of the garage to store the curtain and block edges of light
> - **Modular carpet squares with cable tracks** underneath them
> - Opening ceiling completely and **painting it all black**; put in pipe trussing so we can hang lights, cameras from the ceiling; OR **cable ceiling tracks** so we don't have to worry about zip ties
> - **Sound absorption panels in black placed around ceiling** so they disappear; hung with **wood screws and washers**

I then copied and pasted only the bolded items I was ready to purchase into a separate list below my original notes, and suddenly I had a convenient shopping list I could easily reference while browsing the store.

This example illustrates how even Progressively Summarizing notes from our own conversations can be immensely useful. Often your own thoughts need some distillation before you can take action on them.

> **Derick's home studio recommendations**
>
> - **4 section bifold door** with frosted glass
> - Black theater **blackout curtain** that can be pulled over the entire inside door (to block both light and echoes); **eyelets across the top** so we can hang it or take it off when not in use; OR have a baffle at the corner of the garage to store the curtain and block edges of light
> - **Modular carpet squares with cable tracks** underneath them
> - Opening ceiling completely and **painting it all black**; put in pipe trussing so we can hang lights, cameras from the ceiling; OR **cable ceiling tracks** so we don't have to worry about zip ties
> - **Sound absorption panels in black placed around ceiling** so they disappear; hung with **wood screws and washers**
>
> ---
>
> To buy:
> - ☐ Blackout curtain
> - ☐ Eyelets
> - ☐ Black paint
> - ☐ Cable ceiling tracks

Picasso's Secret: Prune the Good to Surface the Great

We can look at masters of creativity throughout history to see how distilling their ideas shaped their work.

One of Pablo Picasso's most famous drawings, created in 1945 and known as *Picasso's Bull*, offers a master class in how distillation works. It is a sequence of images that he drew to study a bull's essential form. The process of distillation happens in every art form, but this example is unusual in that Picasso preserved each step of his process.

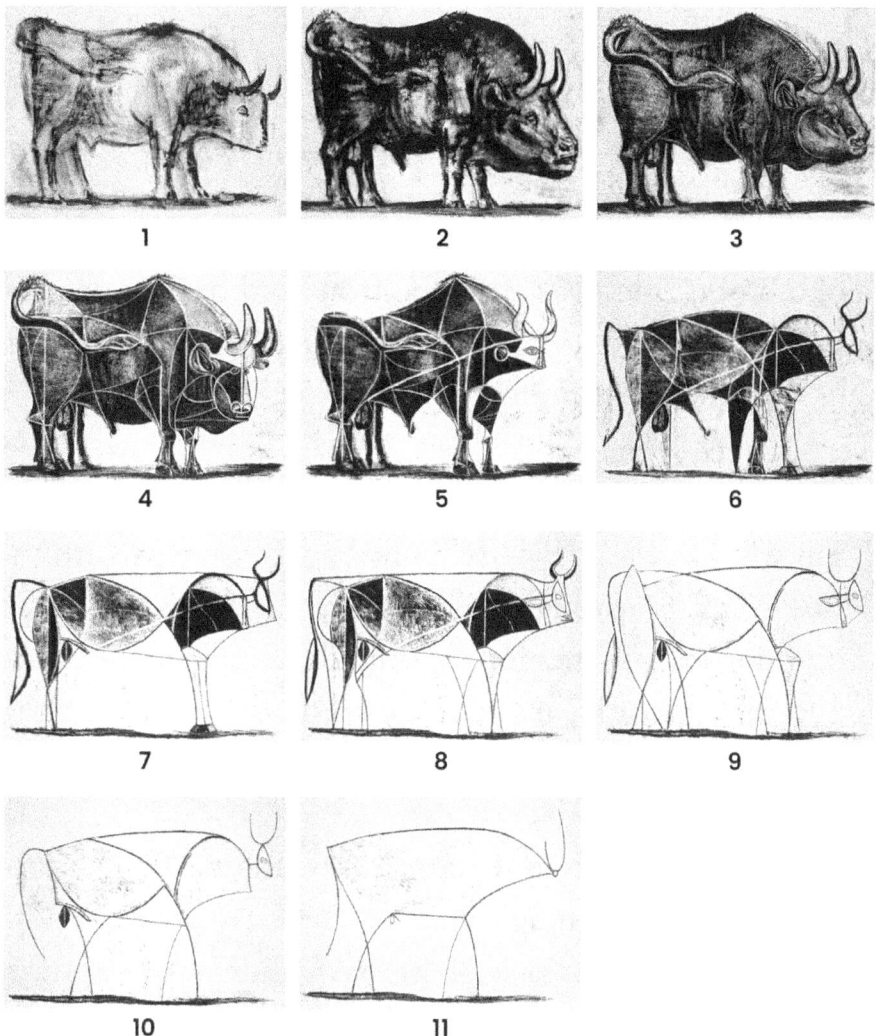

Pablo Picasso, *Le Taureau* (series of 11 lithographs), 1945–46 (© 2021 Estate of Pablo Picasso / Artists Rights Society (ARS), New York).

Starting with the top left image and moving across and down, Picasso deconstructed the shapes of the bull one step at a time. In the first couple of drawings, he adds more detail. The horns are fuller, the tail becomes three-dimensional, and the hide has more depth and texture. Picasso is starting by building up detail so that he has more options to choose from when it comes time to take some away.

The process of distillation begins with the fourth image. He outlines the main muscles of the animal using sharp white lines. Soft curves become more angular, and the animal as a whole starts to take on a more geometric look. In the fifth and sixth images, the drawing starts to become radically simplified as Picasso drops most of the detail in the bull's head and further simplifies its horns, tail, and legs. A thick white line representing the bull's center of gravity is added, cutting across the animal from front to back.

By the last few images, the bull has become nothing more than an interconnected series of simple, black-and-white shapes. The legs have become single lines. Solid blocks of color define the front and back of the animal. In the final drawings, even those details are abstracted away. We end with a drawing that is nothing but a single, continuous stroke, which somehow still manages to capture the very essence of the bull.[6]

Picasso's act of distillation involves stripping away the unnecessary so that only the essential remains. Crucially, Picasso couldn't have started with the single line drawing. He needed to go through each layer of the bull's form step-by-step to absorb the proportions and shapes into his muscle memory. The result points to a mysterious aspect of the creative process: it can end up with a result that looks so simple, it seems like anyone could have made it. That simplicity masks the effort that was needed to get there.

Another example comes from documentary filmmaking. Ken Burns, the renowned creator of award-winning films like *The Civil War*, *Baseball*, and *Jazz*, has said that only a tiny percentage of the raw footage he captures eventually makes it into the final cut. This ratio can be as high as 40- or 50-to-1, which means that for every forty to fifty hours of footage he captures, only one hour makes it into the final film. Along the way, Burns and his team are performing a radical act of distillation—finding the most interesting, surprising, moving moments hidden amidst hundreds of hours of recordings.[VI]

Progressive Summarization is not a method for remembering as much as possible—it is a method for forgetting as much as possible. As you distill your ideas, they naturally improve, because when you drop the merely good parts, the great parts can shine more brightly. To be clear, it takes skill and courage to let the details fall away. As Picasso's bull and Burns's documentaries illustrate, in making decisions about what to keep, we inevitably have to make decisions about what to throw away. You cannot highlight the main takeaways from an article without leaving some points out. You cannot make a highlight reel of a video without cutting some of the footage. You cannot give an effective presentation without leaving out some slides.

The Three Most Common Mistakes of Novice Notetakers

Here are a few guidelines to help you avoid common pitfalls as you embark on highlighting your own notes.

Mistake #1: Over-Highlighting

The biggest mistake people make when they start to distill their notes is that they highlight way too much. You may have experienced this pitfall in school, highlighting paragraph after paragraph or entire pages of textbooks in the vain hope that you'd automatically be able to remember everything in yellow for the test.

When it comes to notetaking for work, less is more. You can capture entire books, articles with dozens of pages, or social media posts by the hundreds. No one will stop you, but you'll quickly learn that such volume will only create a lot more work later on when you have to figure out what all that information means. If you're going to capture everything, you might as well capture nothing.

Remember that notes are not authoritative texts. You don't need to and shouldn't include every tiny detail. They are more like bookmarks peeking out from the pages of a book on the shelf, signaling to you, "Hey! There's something interesting here!" You will always be able to go back and review the full, original source if needed. Your notes only solve the problem of rediscovering those sources when you need them.

A helpful rule of thumb is that each layer of highlighting should include no more than 10–20 percent of the previous layer. If you save a series of excerpts from a book amounting to five hundred words, the bolded second layer should include no more than one hundred words, and highlighted third layer no more than twenty. This isn't an exact science, but if you find yourself highlighting everything, this rule should give you pause.

Mistake #2: Highlighting Without a Purpose in Mind

The most common question I hear about Progressive Summarization is "When should I be doing this highlighting?" The answer is that you should do it *when you're getting ready to create something.*

Unlike Capture and Organize, which take mere seconds, it takes time and effort to distill your notes. If you try to do it with every note up front, you'll quickly be mired in hours of meticulous highlighting with no clear purpose in mind. You can't afford such a giant investment of time without knowing whether it will pay off.

Instead, wait until you know how you'll put the note to use. For example, when I'm preparing to write a blog post or article, I'll usually start by highlighting the most interesting points from a group of notes that I think will be relevant to the topic at hand. That way I have a predictable, not-too-difficult task to get me warmed up for writing, the same way an athlete might have a warm-up and stretching routine.

When I'm about to get on a call with my lawyer, I'll often prepare by highlighting my notes from our last call and drawing out decision points and action items into an agenda. He always thinks I'm well prepared, when in fact I just want to finish the call quickly to minimize the time I'm being billed for!

You have to always assume that, until proven otherwise, any given note won't necessarily ever be useful. You have no idea what your future self will need, want, or be working on. This assumption forces you to be conservative in the time you spend summarizing notes, doing so only when it's virtually guaranteed that it will be worth it.

The rule of thumb to follow is that every time you "touch" a note, you should make it a little more discoverable for your future self[VII]—by adding a highlight, a heading, some bullets, or commentary. This is the "campsite rule" applied to information—leave it better than you found it. This ensures that the notes you interact with most often will naturally become the most discoverable in a virtuous cycle.

Mistake #3: Making Highlighting Difficult

Don't worry about analyzing, interpreting, or categorizing each point to decide whether to highlight it. That is way too taxing and will break the flow of your concentration. Instead, rely on your intuition to tell you when a passage is interesting, counterintuitive, or relevant to your favorite problems or a current project.

Just as you listened for a feeling of internal resonance in deciding what content to save in the first place, the same rule applies for the insights *within* the note. Certain passages will move you, pique your attention, make your heart beat faster, or provoke you. Those are clear signals that you've found something important, and it's time to add a highlight. You can apply the same criteria I introduced earlier in Chapter 4, looking out for individual points that are surprising, useful, inspiring, or personal to decide which ones are worth highlighting.

When you learn the art of distillation, you will gain a lifelong skill that will impact every area of your life. Think of a storyteller who captivates you with every word. Their story is well distilled, with unnecessary details stripped away. Think about the last time you were entranced by a drawing or painting. Its ability to grab you immediately is a sign that the concept behind the artwork is compressed into its most compact form, allowing it to travel efficiently from the canvas straight into your brain.

Even in our daily conversations, the ability to be succinct without missing key details is what leads to exciting conversations that leave both people feeling enlivened. Distillation is at the heart of the communication that is so central to our friendships, our working relationships, and our leadership abilities. Notetaking gives you a way to deliberately practice the skill of distilling every day.

Your Turn: Keep Your Future Self in Mind

The effort we put into Progressive Summarization is meant for one purpose: to make it easy to find and work with our notes in the future.

More is not better when it comes to thinking and creating. Distilling makes our ideas small and compact, so we can load them up into our minds with minimal effort. If you can't locate a piece of information quickly, in a format that's convenient and ready to be put to use, then you might as well not have it at all. Our most scarce resource is time, which means we need to prioritize our ability to quickly rediscover the ideas that we already have in our Second Brain.

When the opportunity arrives to do our best work, it's not the time to start reading books and doing research. You need that research to already be done.[VIII] You can prepare in advance for the future challenges and opportunities you don't even yet know you'll face, by taking advantage of the effort you're already spending reading books, learning new things, and simply being curious about the world around you.

To put what you've just learned into practice immediately, find an interesting piece of content you consumed recently, such as an article, audiobook, or YouTube video. This could be content you've captured already and organized in one of your PARA folders. Or it could be a new piece of content floating around your email inbox or in a read later app.

Start by saving only the best excerpts from that piece of content in a new note, either using copy-paste or a capture tool. This is layer one, the initial excerpts you save in your Second Brain. Next, read through the excerpts, bolding the main points and most important takeaways. Don't make it an analytical decision—listen for a feeling of resonance and let that be your guide for what to bold. These bolded passages are layer two.

Now read through only the bolded passages, and highlight (or, if your notes app doesn't have a highlighting feature, underline) the best of the best passages. The key here is to be very picky: the entire note may have only a few highlighted sentences, or even just one. Not only is that fine, it represents a highly distilled and discoverable note. These highlights are layer three, which is distilled enough for most use cases.

The true test of whether a note you've created is discoverable is whether you can get the gist of it at a glance. Put it aside for a few days and set a reminder to revisit it once you've forgotten most of the details. When you come back to it, give yourself no more than thirty seconds and see if you can rapidly get up to speed on what it's about using the highlights you previously made. You'll quickly be able to tell if you've added too many highlights or too few.

Each time you decide to add a highlight, you are developing your judgment: distinguishing the bits that truly matter from those that don't. This is a skill you can become better at over time. The more you exercise your judgment, the more efficient and enjoyable your notetaking will become because you know that every minute of attention you invest is creating lasting value. There are few things more satisfying than the feeling of making consistent progress.

In the next chapter, we will move on to the final step of CODE, drawing on the material you've collected and distilled and using it to express your own point of view.

I. Discoverability, Wikipedia

II. I like to think of layer one as the "soil"—an excerpt from a source or my own thinking (whether as words, drawings, images, or audio) I initially capture into my notes. They are like the ground on which my understanding will be built. Layer two is "oil," as in "I've struck oil!," conveniently represented by black, bolded text. Layer three is "gold," which is even more valuable, and shines in highlighter yellow in many apps. Layer four is the "gems," the most rare and illuminating finds that I've distilled in my own words as an executive summary.

III. In his book *A New Method of Making Common-Place-Books*, John Locke similarly advised that "We extract only those Things which are Choice and Excellent, either for the Matter itself, or else the Elegancy of the Expression, and not what comes next."

IV. As humans we are exquisitely sensitive to the way information is presented. In web design, a slight change of color for a button or a slightly reworded headline can easily have a double-digit impact on the number of visitors who click on it. Imagine if we put as much thought into how the information on our own devices is presented to us as we do into the public web. Even something as simple as an informative heading, a paragraph break, or a highlighted phrase can make it dramatically easier to absorb a piece of text.

V. Distillation also applies to other kinds of media such as images, audio, and video, but looks quite different and is beyond the scope of this book.

VI. In a class on documentary filmmaking on the educational platform MasterClass, Burns offered his suggestions for keeping track of material related to a film project: "Is there an article in the newspaper that relates to your project? Cut it out and file it. Have you written a draft of narration or dialogue? Print it out and file it. Thought of some great questions to ask your first interview subject? Jot them down on a scrap of paper and file that too."

VII. This principle is called stigmergy—to leave "marks" on the environment that make your future efforts easier. It is a strategy used by ant colonies to find food. If an ant finds a food source, it will bring a piece of it back to the colony, while leaving a special pheromone along the trail. Other ants can follow this trail to find the food for themselves, enabling a multitude of ants to quickly find and collect new sources of food.

VIII. As Sönke Ahrens observes in his book *How to Take Smart Notes*, this is the fundamental paradox at the heart of writing: you have to do the research before you know what you will write about. In his words: "We have to read with a pen in hand, develop ideas on paper and build up an ever-growing pool of externalised thoughts. We will not be guided by a blindly made-up plan picked from our unreliable brains, but by our interest, curiosity and intuition, which is formed and informed by the actual work of reading, thinking, discussing, writing and developing ideas—and is something that continuously grows and reflects our knowledge and understanding externally."

Chapter 7

Express—Show Your Work

Verum ipsum factum ("We only know what we make")
—Giambattista Vico, Italian philosopher

In June 1947, a baby girl named Octavia Estelle Butler was born in Pasadena, California.

Known in her early years as simply "Estelle," she was raised by a single, widowed mother who worked domestic jobs to make ends meet. Painfully shy and introverted from a young age, Estelle became an easy target for bullying at school, which led her to believe she was "ugly and stupid, clumsy, socially hopeless."[1] Her shyness combined with mild dyslexia made schoolwork difficult.

In response, Estelle turned inward to her own imagination and outward to the Pasadena Central Library, where she would spend countless hours reading fairy tales and horse stories, and later, the fantasy and science-fiction novels that would eventually inspire her to become a writer.

Despite the odds stacked against her, this young woman would eventually become one of the most successful and influential science-fiction writers of her generation, winning multiple Hugo and Nebula awards (the genre's highest honors) and in 1995 becoming the first sci-fi writer to receive a MacArthur "Genius" Fellowship.

But Estelle wasn't always so successful. Her teachers at Garfield Elementary School evaluated her earliest writing harshly, with comments like "Hyperbolic" and "You're not even trying" scribbled in the margins.[2] An elementary school teacher once asked, "Why include the science fiction touch? I think the story would be more universal if you kept to the human, earthly touch." The teacher reported to her mother that "She has the understanding, but doesn't apply it. She needs to learn self-discipline."

When she was twelve years old, Estelle watched the 1954 film *Devil Girl From Mars*, a sensationalist B movie that was so terrible it convinced Estelle that she could write something better. She recalls, "Until I began writing my own stories, I never found quite what I was looking for... In desperation, I made up my own."

As the possibility of becoming a professional writer slowly dawned on her, Estelle began her transformation into "Octavia," whom she thought of as her powerful, assertive alter ego. Octavia took on a series of temporary or part-time jobs after graduating from high school: clerical, factory, warehouse, laundry, and food preparation gigs—anything that wasn't too mentally taxing, and that allowed her to maintain her routine of waking before dawn each morning to write.

The emerging Octavia made three rules for herself:

1. Don't leave your home without a notebook, paper scraps, something to write with.

2. Don't walk into the world without your eyes and ears focused and open.

3. Don't make excuses about what you don't have or what you would do if you did, use that energy to "find a way, make a way."

Thus began a lifelong relationship with her commonplace books. Butler would scrape together twenty-five cents to buy small Mead memo pads, and in those pages she took notes on every aspect of her life: grocery and clothes shopping lists, last-minute to-dos, wishes and intentions, and calculations of her remaining funds for rent, food, and utilities. She meticulously tracked her daily writing goals and page counts, lists of her failings and desired personal qualities, her wishes and dreams for the future, and contracts she would sign with herself each day for how many words she committed to write.

Of course, Butler also gathered material for her fantastic stories: lyrics to songs she'd heard on the radio, an idea for a character's name or motivation, a new topic to research, details of news stories—everything she needed to build the worlds her stories would take place in. She studied dozens of topics—anthropology, English, journalism, and speech. She traveled to the Amazon and Incan ruins in Peru to get a firsthand taste of biodiversity and civilizational collapse. Like a journalist, Butler had a love for cold, hard facts to imbue her stories with a sense of authenticity and concreteness: "The greater your ignorance the more verifiably accurate must be your facts," she once remarked.

One of Butler's novels, *The Parable of the Sower*, hit the *New York Times* bestseller list for the first time in 2020,[3] fulfilling one of Butler's life goals fourteen years after her death. The book portrays a postapocalyptic future in the aftermath of runaway climate disasters, in which small communities must band together in order to survive. These eerily prescient forecasts resonated with readers as the COVID-19 pandemic unfolded, as our own time began to seem similarly bleak and uncertain. The radical reimagination of what life could look like in the midst of a crisis was no longer idle speculation—it had become a daily preoccupation for people around the world. Butler has been called a prophet for her ability to forecast the future, but she always said that her work came from simply imagining, "If this goes on... it extrapolates from current technology, current ecological conditions, current social conditions, current practices of any sort. It offers good possibilities—as well as warnings."

Butler knew that science fiction was more than entertainment. It was a transformative way of viewing the future. As one of the first Black women to gain recognition in the sci-fi genre, Butler explored ideas and themes that had been previously ignored: the potential consequences of environmental collapse due to climate change, corporate greed and the growing gap between the wealthy and poor, gender fluidity and the "othering" of marginalized groups, and criticism of the hierarchical nature of society, among other themes.

Butler pioneered Afrofuturism, a genre that cast African Americans as protagonists who embrace radical change in order to survive. Her stories allowed her readers to visualize futures in which marginalized people are heroes, not victims. Through her writing, she expanded our vision of the future to include the untold stories of the disenfranchised, the outcast, and the unconventional.

How do we know so much about even the tiniest details of Butler's life? Because she kept it all—journals, commonplace books, speeches, library call slips, essay and story drafts, school notes, calendars, and datebooks as well as assorted odds and ends like school progress reports, bus passes, yearbooks, and contracts. This collection contained 9,062 items and filled 386 boxes when it was donated to the Huntington Library in San Marino, California, after Butler's passing.[4]

How could a painfully shy little girl become a world-renowned, award-winning writer? How could an impoverished and overworked young woman emerge as a powerful prophet of the future? In her own words: "My mother was a maid, my father shined shoes, and I wanted to write science fiction, who was I kidding?"

Butler did it by drawing on her life experience: "The painful, horrifying, the unpleasant things that happen, affect my work more strongly than the pleasant ones. They're more memorable and more likely to goad me into writing interesting stories."

She used her notes and her writing to confront her demons: "The biggest obstacle I had to overcome was my own fear and self-doubt—fear that maybe my work really wasn't good enough, maybe I wasn't smart enough; maybe the people telling me I couldn't make it were right."

She used every bit of insight and detail she could muster from both her daily life and the books she immersed herself in: "Use what you have; even if it seems meager, it may be magic in your hands." Butler found a way to express her voice and her ideas even when her circumstances made it seem impossible.

The myth of the writer sitting down before a completely blank page, or the artist at a completely blank canvas, is just that—a myth. Professional creatives constantly draw on outside sources of inspiration—their own experiences and observations, lessons gleaned from successes and failures alike, and the ideas of others. If there is a secret to creativity, it is that it *emerges* from everyday efforts to gather and organize our influences.

How to Protect Your Most Precious Resource

As knowledge workers, attention is our most scarce and precious resource.

The creative process is fueled by attention at every step. It is the lens that allows us to make sense of what's happening, to notice what resources we have at our disposal, and to see the contribution we can make. The ability to intentionally and strategically allocate our attention is a competitive advantage in a distracted world. We have to jealously guard it like a valuable treasure.

You have twenty-four hours in a day, but how many of those hours include your highest-quality attention? Some days are so frenetic and fragmented that you might not have any at all. Attention can be cultivated but also destroyed—by distractions, interruptions, and environments that don't protect it. The challenge we face in building a Second Brain is how to establish a system for personal knowledge that *frees up* attention, instead of taking more of it.

We've been taught that it's important to work "with the end in mind." We are told that it is our responsibility to deliver results, whether that is a finished product on store shelves, a speech delivered at an event, or a published technical document.

This is generally good advice, but there is a flaw in focusing only on the final results: all the intermediate work—the notes, the drafts, the outlines, the feedback—tends to be underappreciated and undervalued. The precious attention we invested in producing that in-between work gets thrown away, never to be used again. Because we manage most of our "work-in-process" in our head, as soon as we finish the project and step away from our desks, all that valuable knowledge we worked so hard to acquire dissolves from our memory like a sandcastle washed away by the ocean waves.

If we consider the focused application of our attention to be our greatest asset as knowledge workers, we can no longer afford to let that intermediate work disappear. If we consider how precious little time we have to produce something extraordinary in our careers, it becomes imperative that we recycle that knowledge back into a system where it can become useful again.

What are the knowledge assets you're creating today that will be most reusable in the future? What are the building blocks that will move forward your projects tomorrow? How can you package up what you know in a form that you'll be able to revisit it again and again no matter what endeavors you take on in the future?

The final stage of the creative process, Express, is about refusing to wait until you have everything perfectly ready before you share what you know. It is about expressing your ideas earlier, more frequently, and in smaller chunks to test what works and gather feedback from others. That feedback in turn gets drawn in to your Second Brain, where it becomes the starting point for the next iteration of your work.

Intermediate Packets: The Power of Thinking Small

The idea of dividing our work into smaller units isn't new. You've probably heard this advice a hundred times: if you're stuck on a task, break it down into smaller steps.

Every profession and creative medium has its own version of "intermediate steps" on the way to full-fledged final works. For example:

- "Modules" in software development
- "Betas" tested by start-ups
- "Sketches" in architecture
- "Pilots" for television series
- "Prototypes" made by engineers
- "Concept cars" in auto design
- "Demos" in music recording

Each of these terms is the equivalent of a "rough draft" you create as part of the process of making something new.

Here's what most people miss: it's not enough to simply divide tasks into smaller pieces—you then need a *system* for managing those pieces. Otherwise, you're just creating a lot of extra work for yourself trying to keep track of them.

That system is your Second Brain, and the small pieces of work-in-process it contains I call "Intermediate Packets." Intermediate Packets are the concrete, individual building blocks that make up your work.[1] For example, a set of notes from a team meeting, a list of relevant research findings, a brainstorm with collaborators, a slide deck analyzing the market, or a list of action items from a conference call. Any note can potentially be used as an Intermediate Packet in some larger project or goal.

Think of a salesperson planning a new campaign for a health-branded energy drink. Sales might seem like the kind of job that is least related to "knowledge management." Isn't it all about making calls, having meetings, sending pitches, and closing deals?

If we take a closer look, there are many building blocks such a sales job relies on. The company brochure, the sales prospectus, the cold-calling scripts, the list of warm leads, notes from past calls with an important distributor—these are the assets that a salesperson depends on for their performance.

Like LEGO blocks, the more pieces you have, the easier it is to build something interesting. Imagine that instead of starting your next project with a blank slate, you started with a set of building blocks—research findings, web clippings, PDF highlights, book notes, back-of-the-envelope sketches—that represent your long-term effort to make sense of your field, your industry, and the world at large.

Our time and attention are scarce, and it's time we treated the things we invest in—reports, deliverables, plans, pieces of writing, graphics, slides—as knowledge assets that can be reused instead of reproducing them from scratch. Reusing Intermediate Packets of work frees up our attention for higher-order, more creative thinking. Thinking *small* is the best way to elevate your horizons and expand your ambitions.

There are five kinds of Intermediate Packets you can create and reuse in your work:

- **Distilled notes:** Books or articles you've read and distilled so it's easy to get the gist of what they contain (using the Progressive Summarization technique you learned in the previous chapter, for example).

- **Outtakes:** The material or ideas that didn't make it into a past project but could be used in future ones.

- **Work-in-process:** The documents, graphics, agendas, or plans you produced during past projects.

- **Final deliverables:** Concrete pieces of work you've delivered as part of past projects, which could become components of something new.

- **Documents created by others:** Knowledge assets created by people on your team, contractors or consultants, or even clients or customers, that you can reference and incorporate into your work.

If you're reading how-to articles in your free time, you can save the best tips in your notes and turn them into *distilled notes* for when it's time to put them to use. If you're writing an essay and decide to cut a paragraph, you can save those *outtakes* in case you ever write a follow-up. If you are in product development and create a detailed set of requirements, you can save that *work-in-process* as a template for future products. If you're a management consultant, you can save the slides you presented to an executive team as a *final deliverable*, and reuse them for similar presentations. If you're a lab scientist and a colleague designs the perfect lab protocol, you can reuse and improve that *document* for your own use (with their permission, of course).

You should always cite your sources and give credit where credit is due. A scientist doesn't obscure her sources—she points to them so others can retrace her footsteps. We all stand on the shoulders of giants, and it's smart to build on the thinking they've done rather than try to reinvent the wheel.

Making the shift to working in terms of Intermediate Packets unlocks several very powerful benefits.

First, you'll become interruption-proof because you are focusing only on one small packet at a time, instead of trying to load up the entire project into your mind at once. You become less vulnerable to interruptions, because you're not trying to manage all the work-in-process in your head.

Second, you'll be able to make progress in any span of time. Instead of waiting until you have multiple uninterrupted hours—which, let's face it, is rare and getting rarer—you can look at how many minutes you have free and choose to work on an IP that you can get done within that time, even if it's tiny. Big projects and goals become less intimidating because you can just keep breaking them down into smaller and smaller pieces, until they fit right into the gaps in your day.

Third, Intermediate Packets increase the quality of your work by allowing you to collect feedback more often. Instead of laboring for weeks in isolation, only to present your results to your boss or client and find out you went in the wrong direction, you craft just one small building block at a time and get outside input before moving forward. You'll find that people give much better feedback if they're included early, and the work is clearly in progress.

Fourth, and best of all, eventually you'll have so many IPs at your disposal that you can execute entire projects just by assembling previously created IPs. This is a magical experience that will completely change how you view productivity. The idea of starting anything from scratch will become foreign to you—why not draw on the wealth of assets you've invested in in the past? People will marvel at how you're able to deliver at such a high standard so consistently. They'll wonder how you find the time to do so much careful thinking, when in fact you're not working harder or longer—all you're doing is drawing on a growing library of Intermediate Packets stored in your Second Brain. If they are truly valuable assets, then they deserve to be managed, just like any other asset you possess.

Intermediate Packets are really a new lens through which you can perceive the atomic units that make up everything you do. By "thinking small," you can focus on creating just one IP each time you sit down to work, without worrying about how viable it is or whether it will be used in the exact way you envisioned. This lens reframes creativity as an ongoing, continual cycle of delivering value in small bits, rather than a massive all-consuming endeavor that weighs on you for months.

Assembling Building Blocks: The Secret to Frictionless Output

Every time you make a sketch, design a slide, record a short video on your phone, or post on social media, you are undertaking a small creative act that produces a tangible by-product. Consider the different kinds of documents and other content that you probably regularly produce as part of your normal routines:

- Favorites or bookmarks saved from the web or social media
- Journal or diary entries with your personal reflections
- Highlights or underlined passages in books or articles
- Messages, photos, or videos posted on social media
- Slides or charts included in presentations
- Diagrams, mind maps, or other visuals on paper or in apps
- Recordings of meetings, interviews, talks, or presentations
- Answers to common questions you receive via email
- Written works, such as blog posts or white papers
- Documented plans and processes such as agendas, checklists, templates, or project retrospectives

While you can sit down to purposefully create an IP, it is far more powerful to simply *notice* the IPs that you have already produced and then to take an extra moment to save them in your Second Brain.

Let's look at an example: planning a large conference. If it's a brand-new event, or you've never organized a conference before, it might seem like you have to produce everything from scratch. However, if you break down that mega-project into concrete chunks, suddenly the components that you'll need become clear:

- A conference agenda
- A list of interesting breakout sessions
- A checklist for streaming the keynote sessions
- An email announcing the conference to your network
- An invitation for people to be speakers or panelists
- A conference website

These are some of the building blocks that you'll need to be able to run the conference. You could put them all on your to-do list and make them yourself, but there's a different, much faster, and more effective approach. Ask yourself: How could you *acquire* or *assemble* each of these components, instead of having to make them yourself?

The conference agenda could easily be modeled on an agenda from a different conference, with the topics and speaker names switched out. You could start compiling a list of potential breakout sessions, adding any topic suggested by others that strikes you as interesting. You might have a checklist for delivering effective keynotes left over from a live event you've organized in the past. Emails can draw on an archive of examples you've saved from other conferences you've attended. Screenshots of conference websites you admire are the best possible starting point for designing your own.

Our creativity thrives on examples. When we have a template to fill in, our ideas are channeled into useful forms instead of splattered around haphazardly. There are best practices and plentiful models for almost anything you might want to make.

Most professionals I work with already have and use Intermediate Packets—that's the point! Your Second Brain is the repository of things you are *already* creating and using anyway. All we are doing is adding a little bit of structure and intentionality to how we use them: capturing them in one place, such as a digital notes app, so we can find them with a search; organizing them according to our projects, areas, and resources, so we have a dedicated place for each important aspect of our lives; and distilling them down to their most essential points, so they can be quickly accessed and retrieved.

Once we've completed these initial steps, expression transforms from a gut-wrenching, agonizing feat to a straightforward assembly of existing packets of work.

Over time, your ability to quickly tap these creative assets and combine them into something new will make all the difference in your career trajectory, business growth, and even quality of life. In the short term, it might not matter. You might be able to scramble and put together a particular document right when you need it, but there will be a slowly accumulating, invisible cost. The cost of not being quite sure whether you have what you need. The stress of wondering whether you've already completed a task before. There is a cost to your sleep, your peace of mind, and your time with family when the full burden of constantly coming up with good ideas rests solely on your fickle biological brain.

How to Resurface and Reuse Your Past Work

The Express step is where we practice and hone our ability to retrieve what we need, when we need it. It's the step where we build the confidence that our Second Brain is working for us.

Let's take a closer look at the process of retrieval: How can you find and retrieve Intermediate Packets when you need them?

This isn't a trivial question, because the connection between IPs we've saved in the past and future projects is often quite unpredictable. A concert poster on the side of a building you snapped a photo of might inform the shapes in a logo you're designing. A song overheard on the subway might influence a jingle you're writing for your child's school play. An idea about persuasion you read in a book might become a central pillar in a health campaign you are organizing for your company.

These are some of the most valuable connections—when an idea crosses the boundaries between subjects. They can't be planned or predicted. They can emerge only when many kinds of ideas in different shapes and sizes are mixed together.

This inherent unpredictability means that there is no single, perfectly reliable retrieval system for the ideas contained in your notes. Instead, there are four methods for retrieval that overlap and complement one another. Together they are more powerful than any computer yet more flexible than any human mind. You can step through them in order until you find what you're looking for.

Those four retrieval methods are:

1. Search
2. Browsing

3. Tags
4. Serendipity

Retrieval Method #1: Search

The search function in your notes app is incredibly powerful. The same technology that has revolutionized how we navigate the web via search engines is also useful for navigating our private knowledge collections.

Search has the benefit of costing almost nothing in terms of time and effort. Just by saving your notes in a central place, you enable software to search their full contents in seconds. You can run multiple searches in quick succession, running down rabbit trails through your knowledge garden as you try out different variations of terms.

This quick, iterative approach to searching is where notes apps shine—you don't have to open and close individual notes one at a time as in traditional word processing. In a sense, every note in your Second Brain is already "open," and you can view or interact with its contents with a mere click or tap.

Search should be the first retrieval method you turn to. It is most useful when you already know more or less what you're looking for, when you don't have notes saved in a preexisting folder, or when you're looking for text, but as with every tool, it has its limitations. If you don't know exactly what you're looking for, don't have a preexisting folder to look through, or are interested in images or graphics, it's time to turn to browsing.

Retrieval Method #2: Browsing

If you've followed the PARA system outlined in Chapter 5 to organize your notes, you already have a series of dedicated folders for each of your active projects, areas of responsibility, resources, and archives.

Each of these folders is a dedicated environment designed specifically for focusing on that domain of your life. Each one can contain a wide range of content, from brief notes dashed off during a phone call to polished Intermediate Packets that you've already used in past projects. When the time comes to take action, you'll be able to enter the appropriate workspace and know that everything found there is relevant to the task at hand.

As powerful as search can be, studies[5] have found that in many situations people strongly prefer to navigate their file systems manually, scanning for the information they're looking for. Manual navigation gives people control over how they navigate, with folders and file names providing small contextual clues about where to look next.[6] Browsing allows us to gradually home in on the information we are looking for, starting with the general and getting more and more specific. This kind of browsing uses older parts of the brain that developed to navigate physical environments, and thus comes to us more naturally.[11]

There are a variety of features offered by notes apps that make it easy to browse your hierarchy of folders. Some apps allow you to "sort" a list of notes by different criteria, such as date created. This gives you an interactive timeline of your ideas from newest to oldest. Other apps allow you to show only images and web clippings, enabling rapid visual scanning to see if anything catches your eye. Most notes apps allow you to open multiple windows and compare their contents side by side so you can look for patterns and move content between them.

Once again, there are limitations to what you can find by browsing folders. Sometimes you know a project is coming and can start saving things to a project folder in advance, but sometimes you don't. Sometimes it's very clear which area of your business a note is related to, but often you have no idea where to put it. Many notes end up being useful in completely unexpected ways. We want to encourage that kind of serendipity, not fight it!

It is for the unforeseen and the unexpected that tags really shine.

Retrieval Method #3: Tags

Tags are like small labels you can apply to certain notes regardless of where they are located. Once they are tagged, you can perform a search and see all those notes together in one place. The main weakness of folders is that ideas can get siloed from each other, making it hard to spark interesting connections. Tags can overcome this limitation by infusing your Second Brain with connections, making it easier to see cross-disciplinary themes and patterns that defy simple categorization.

For example, maybe you work in customer service and notice that the same questions from customers keep coming up again and again. You might decide to write a Frequently Asked Questions page and add it to your company website. That is a project, but not one that you previously recognized and started gathering materials for. You might have various notes that you want to draw on to design this page, but don't want to move them from the project, area, and resource folders where they're currently located.

It's time for tags. You could take fifteen minutes and perform a series of searches for terms relevant to the FAQs you'll be writing. For any useful note you find, apply a tag called "FAQ" while leaving it right where you found it. Once you've found enough material to work with, you can perform a single search—for the "FAQ" tag—and instantly see all the notes you've tagged collected in one place. Now you are free to review them more closely, highlight any specific points you want to use, and move those points into an outline to guide your writing.

I don't recommend using tags as your primary organizational system. It takes far too much energy to apply tags to every single note compared to the ease of searching with keywords or browsing your folders. However, tags can come in handy in specific situations when the two previous retrieval methods aren't up to the task, and you want to spontaneously gather, connect, and synthesize groups of notes on the fly.[III]

Retrieval Method #4: Serendipity

The fourth retrieval method is the most mysterious but, in many ways, the most powerful. Beyond searching, browsing, and tagging, there is a frontier of possibility that simply cannot be planned or predicted by human minds. There are moments when it feels like the stars align and a connection between ideas jumps out at you like a bolt of lightning from a blue sky. These are the moments creatives live for.

There is no way to plan for them, but that doesn't mean we can't create the ideal conditions for them to arise. This is the main reason we put all sorts of different kinds of material, on many subjects and in diverse formats, all jumbled together in our Second Brain. We are creating a soup of creative DNA to maximize the chance that new life emerges.

Serendipity takes a few different forms when it comes to retrieval.

First, while using the previous retrieval methods, it is a good idea to keep your focus a little broad. Don't begin and end your search with only the specific folder that matches your criteria. Make sure to look through related categories, such as similar projects, relevant areas, and different kinds of resources.

When starting a project, I'll often look at five or six PARA folders just in case they contain something useful. Since you've carefully curated the contents of these folders, you won't be faced with too much material in any given one. If you're using Progressive Summarization to distill your notes, as explained in Chapter 6, you can focus only on the highlighted passages and review notes far faster than having to read every word. It usually takes me less than thirty seconds on average to review a highlighted note, which means I can set aside just ten minutes and review twenty of them or more.

Second, serendipity is amplified by visual patterns. This is why I strongly suggest saving not only text notes but images as well (which is difficult to do in other kinds of software such as word processors). Our brains are naturally attuned to visuals. We intuitively absorb colors and shapes in the blink of an eye, using far less energy than it takes to read words. Some digital notes apps allow you to display only the images saved in your notes, which is a powerful way of activating the more intuitive, visual parts of your brain.

Third, sharing our ideas with others introduces a major element of serendipity. When you present an idea to another person, their reaction is inherently unpredictable. They will often be completely uninterested in an aspect you think is utterly fascinating; they aren't necessarily right or wrong, but you can use that feedback either way. The reverse can also happen. You might think something is obvious, while they find it mind-blowing. That is also useful feedback. Others might point out aspects of an idea you never considered, suggest looking at sources you never knew existed, or contribute their own ideas to make it better. All these forms of feedback are ways of drawing on not only your first and Second Brains, but the brains of others as well.

Three Stages of Expressing: What Does It Look Like to Show Our Work?

In Chapter 3, I explained how people tend to move through three distinct stages as they grow their Second Brain and refine their knowledge management skills—remembering, connecting, and creating.

Let's look at examples of how each of these can play out using case studies from former students of mine.

Remember: Retrieve an Idea Exactly When It's Needed

Benigno is a father and business consultant in the Philippines, and one of his goals for building a Second Brain was to better understand the emerging cryptocurrency trend. He had tried other organizing methods before but found that the information he collected was always difficult to access. Time and again, he would "keep reading and bookmarking, then forget about it."

Benigno came across an article on an innovative new kind of cryptocurrency and took a few minutes to save some excerpts from it in his notes. When a few of his friends became interested in the topic, he took eight minutes to progressively summarize the best excerpts before sharing the summarized article with them. The time that he had spent reading and understanding a complex subject paid off in time savings for his friends, while also giving them a new interest to connect over.

In Benigno's words, "I instinctively knew that just sending a long article to friends usually doesn't do anything, but because the text I am sending has been highlighted they can do a quick scan of it. Also I now have material for a future article I have been planning… all thanks to CODE."

You don't need to invent a new theory or write the next great novel to derive value from your Second Brain. Within days of capturing ideas that resonate with you, you'll start to notice opportunities to share them with others to their benefit.

Connect: Use Notes to Tell a Bigger Story

Patrick is a church pastor in Colorado, and he uses his Second Brain to help him design memorial services, which for him are a deeply creative experience about honoring life.

His goal with creating a memorial service is to "tell the story of someone's life in a way that honors and makes meaning in retrospect about how their life unfolded." In the past, this entailed a heavy lift. But by seeing his services through the lens of his Second Brain, Patrick realized that his job was simply to gather and connect some of the themes and stories he heard in a way that was meaningful to the deceased person's loved ones.

Patrick used this realization to change his creative process. He began recording his conversations using an automatic transcription app on his smartphone, allowing him to be fully present with grieving families knowing he had every word captured. He saved all the transcripts from his conversations, plus obituaries, photos, and other related documents into the PARA project folder for each memorial, so he could see it all in one place. Instead of spending five to seven hours at the end of all his interviews to distill everything he had heard, he began spending fifteen minutes after each interview to highlight only the parts that resonated.

In Patrick's words: "Using my first brain only for what it is best at is freedom. Freedom to be present and not multitasking as I sit with grieving people hearing stories of their loved ones. Freedom to know I have everything recorded. Freedom to know that when I go to pull the memorial together, 80 percent of the work is already done."[IV]

Creative expression isn't always about self-promotion or advancing our own career. Some of our most beautiful, creative acts are ones in which we connect the dots for others in ways they wouldn't be able to do themselves.

Create: Complete Projects and Accomplish Goals Stress-Free

Rebecca is a professor of educational psychology at a university in Florida, and she uses her digital notes to create programs and presentations as part of her teaching.

Before building her Second Brain, Rebecca would wait until she had a large block of time available to put together her ideas for a talk. As a busy professional and mother, those large uninterrupted blocks of time seemed scarcer than ever.

Her digital notes gave her another way of making progress. In the weeks leading up to an event, on days she felt inspired, Rebecca began dropping short notes into her notes inbox with ideas of things she might want to include. By the time she sat down to write the outline, she realized she already had all the Intermediate Packets—metaphors, research facts, stories, diagrams—she needed at her fingertips. All she had to do was string together the notes and existing IPs she had already captured.

In Rebecca's words: "I'm able to look at my priorities—my priorities for my work, my priorities for my family, my marriage, etc.—and then only focus in the moment on those projects that are on my plate now."

Whatever you are responsible for creating—whether it is documents or presentations or decisions or outcomes—your Second Brain is a vital repository of all the bits and pieces you'll want in front of you when you sit down to focus. It is a creative environment you can step into at any time, in any place, when it's time to make things happen.

Creativity Is Inherently Collaborative

A common myth of creativity is that of the solitary artist, working in total isolation. We are implicitly told that we must shut ourselves off from the influence of others and flesh out our masterpiece by the sweat of our brow.

In my experience, this isn't how creativity works at all. It doesn't matter what medium you work in; sooner or later you must work with others. If you're a musician, you'll need a sound engineer to mix the record. If you're an actor, you'll need a director who believes in you. Even writing a book, which may suggest images of a lonely cabin deep in the woods, is an intensively social exercise. A book is created out of a dance between an author and their editor.

Reframing your work in terms of Intermediate Packets isn't just about doing the same old stuff in smaller chunks. That doesn't unlock your true potential. The transformation comes from the fact that smaller chunks are inherently more shareable and collaborative.

It is much easier to show someone a small thing, and ask for their thoughts on it, rather than the entire opus you're creating. It's less confronting to hear criticism on one small aspect of your work, at an early stage when you still have time to correct it, than getting a negative reaction after months of effort. You can use each little piece of intermediate feedback to refine what you're making—to make it more focused, more appealing, more succinct, or easier to understand.

The fundamental difficulty of creative work is that we are often too close to it to see it objectively. Getting feedback is really about borrowing someone else's eyes to see what only a novice can see. It's about stepping outside your subjective point of view and noticing what's missing from what you've made.

Once you understand how incredibly valuable feedback is, you start to crave as much of it as you can find. You start looking for every opportunity to share your outputs and gain some clarity on how other people are likely to receive it. These moments are so important that you will begin changing how you work in order to get feedback as early and often as possible, because you know it is much easier to gather and synthesize the thoughts of others than to come up with an endless series of brilliant thoughts on your own. You will begin to see yourself as the curator of the collective thinking of your network, rather than the sole originator of ideas.

There are certain notes in your Second Brain that are disproportionately valuable, that you will find yourself returning to again and again. These are the cornerstones of your work on which everything else is built, but you can't usually know which notes are cornerstones up front. You discover them by sharing your ideas with others, and seeing which ones resonate with them. It is by sharing our ideas with other people that we discover which ones represent our most valuable expertise.

Everything Is a Remix

The CODE Method is based on an important aspect of creativity: that it is always a remix of existing parts. We all stand on the shoulders of our predecessors. No one creates anything out of a pure void.

Kitbashing is a practice used in making small-scale models for action movies like *Star Wars* and *Indiana Jones*. To stay within time and budget, model makers buy prefabricated commercial kits and recombine them into new models for their sets. Instead of fabricating new pieces from scratch, these premade parts—from models of World War II flak cannons, US Navy battleships, fighter planes, T-34 tanks, and submarines—can be used to add texture and fine detail to special effects scenes in a fraction of the time and at a fraction of the cost it would normally take.

Adam Savage, host of the popular *MythBusters* television show and a skilled model maker, noted that because these pieces are so versatile, "There were some kits to which you would return repeatedly." As part of the team at Industrial Light & Magic, the studio behind many special effects films, he relied on a particular kit that found its way into almost every model the team ever built.[7]

Don't take the work of others wholesale; borrow *aspects* or *parts* of their work. The shape of a banner on a web page, the layout of a slide, the style of a song—these are like the ingredients you put in a blender before hitting the button and mixing it into your own recipe. Of course, cite all your sources and influences, even if you don't strictly have to. Giving credit where credit is due doesn't lessen the value of your contribution—it increases it. Having a Second Brain where all your sources are clearly documented will make it much easier to track them down and include those citations in the finished version.

I remember the first time someone referred to what I do as "your work." It dawned on me that I had a body of work that stood on its own, that had an identity distinct from mine. This is a turning point in the life of any creative professional—when you begin to think of "your work" as something separate from yourself.

Reframing your productivity in terms of Intermediate Packets is a major step toward this turning point. Instead of thinking of your job in terms of *tasks*, which always require you to be there, personally, doing everything yourself, you will start to think in terms of *assets* and *building blocks* that you can assemble.

As the potential of your intellectual assets becomes apparent, you'll start to look for any way to spend your time creating such assets and avoid one-off tasks whenever possible. You will start to seek out ways of acquiring or outsourcing the creation of these assets to others, instead of assuming you have to build them all yourself. These changes will enable you to get things done at a pace that is far beyond what mere "productivity tips" can ever achieve.

Even if you're not writing a book now, or creating a presentation now, or developing a new framework now, that doesn't mean you never will. Every little digital artifact you create—the emails, the meeting notes, the project plans, the templates, the examples—is part of the ongoing evolution of your body of work. They are like the neurons in an intelligent organism that is growing, evolving, reaching for higher levels of consciousness with each new experience it has.

Your Turn: You Only Know What You Make

My favorite quote about creativity is from the eighteenth-century philosopher Giambattista Vico: *Verum ipsum factum*. Translated to English, it means "We only know what we make."

To truly "know" something, it's not enough to read about it in a book. Ideas are merely thoughts until you put them into action. Thoughts are fleeting, quickly fading as time passes. To truly make an idea stick, you have to engage with it. You have to get your hands dirty and apply that knowledge to a practical problem. We learn by making concrete things—before we feel ready, before we have it completely figured out, and before we know where it's going.

It is when you begin expressing your ideas and turning your knowledge into action that life really begins to change. You'll read differently, becoming more focused on the parts most relevant to the argument you're building. You'll ask sharper questions, no longer satisfied with vague explanations or leaps in logic. You'll naturally seek venues to show your work, since the feedback you receive will propel your thinking forward like nothing else. You'll begin to act more deliberately in your career or business, thinking several steps beyond what you're consuming to consider its ultimate potential.

It's not necessarily about becoming a professional artist, online influencer, or business mogul: it's about taking ownership of your work, your ideas, and your potential to contribute in whatever arena you find yourself in. It doesn't matter how impressive or grand your output is, or how many people see it. It could be just between your family or friends, among your colleagues and team, with your neighbors or schoolmates—what matters is that you are finding your voice and insisting that what you have to say matters. You have to value your ideas enough to share them. You have to believe that the smallest idea has the potential to change people's lives. If you don't believe that now, start with the smallest project you can think of to begin to prove to yourself that your ideas can make a difference.

You might realize you have lots of notes on eating healthy and decide to experiment with your own take on a classic recipe. You might see the notes from courses you've taken to improve your project management skills and decide to put them together into a presentation for your coworkers. You could draw on the insights and life experiences you've written about in your notes to write a blog post or record a YouTube video to help people who are facing a similar challenge.

All of these are acts of self-expression enabling you to begin unlocking your full creative potential.

I. Intermediate Packets are abbreviated as IPs, a lucky coincidence that is appropriate, because they are absolutely your Intellectual Property. You created them, you own them, and you have the right to use them again and again in any future project.

II. Barbara Tversky, a professor of psychology and education at Teachers College in New York, notes that "We are far better and more experienced at spatial thinking than at abstract thinking. Abstract thought can be difficult in and of itself, but fortunately it can often be mapped onto spatial thought in one way or another. That way, spatial thinking can substitute for and scaffold abstract thought."

III. Tagging for personal knowledge management is a subject unto itself. While not necessary to get started, I've written a free bonus chapter on tags you can download at buildingasecondbrain.com/bonuschapter.

IV. One of my favorite rules of thumb is to "Only start projects that are already 80 percent done." That might seem like a paradox, but committing to finish projects only when I've already done most of the work to capture, organize, and distill the relevant material means I never run the risk of starting something I can't finish.

PART THREE

The Shift
Making Things Happen

Chapter 8

The Art of Creative Execution

Creative products are always shiny and new; the creative process is ancient and unchanging.
—Silvano Arieti, psychiatrist and author of *Creativity: The Magic Synthesis*

I was fortunate to grow up in a multicultural household full of art and music.

My mother is a singer and guitarist from Brazil, and some of my earliest memories are of her soprano voice singing beautiful Portuguese lyrics to the tune of a classical guitar. My father is a professional painter born in the Philippines. His canvases bursting with colorful fruits, verdant landscapes, and monumental figures covered every wall of our house, giving our home the ambience of an art gallery.

I've never recognized the common stereotypes of the "tortured artist"—mercurial, unpredictable, brooding, and unreliable. My father is one of the most orderly, responsible people I've ever met. Yet this regularity didn't take away from his fantastically creative artwork—it contributed to it. I saw how rigorous his routines had to be to allow him to pursue his creative calling while raising a family.

He had a series of what he called his "strategies." These were habits and tricks that he used to integrate creativity into every aspect of his life and quickly get into a creative state of mind whenever he had time to paint.

During sermons at our local church, my father would practice sketching biblical stories in a small paper notebook as he listened. Those sketches would often become the starting point for larger, full-scale works measuring eight or ten feet high. While browsing the supermarket he would buy vegetables with unusual shapes to take home and incorporate into his still lifes. Our groceries doubled as models before we ate them. Often in the evening when we were watching TV together as a family, I'd catch him looking off to the side, at the wall of our living room, where he had hung a painting he was working on. He said he could have insights about what was missing by looking at it in a new light and out of the corner of his eye.

My father planned for creativity. He strategized his creativity. When it was time to make progress on a painting, he gave it his full focus, but that wasn't the only time he exercised his imagination. Much of the rest of the time he was collecting, sifting through, reflecting on, and recombining raw material from his daily life so that when it came time to create, he had more than enough raw material to work with. This attention to organizing his creative influences fueled a prolific body of work made up of thousands of paintings created over decades, while still allowing him plenty of time to attend our soccer games, cook delicious meals, and travel widely as a family.

What I learned from my father is that by the time you sit down to make progress on something, all the work to gather and organize the source material needs to already be done. We can't expect ourselves to instantly come up with brilliant ideas on demand. I learned that innovation and problem-solving depend on a routine that systematically brings interesting ideas to the surface of our awareness.[I]

All the steps of the CODE Method are designed to do one thing: to help you put your digital tools to work for you so that your human, fallible, endlessly creative first brain can do what it does best. Imagine. Invent. Innovate. Create.

Building a Second Brain is really about *standardizing* the way we work, because we only really improve when we standardize the way we do something. To get stronger, you need to lift weights using the correct form. A musician relies on standardized notes and time signatures so they don't have to reinvent the basics from scratch every time. To improve your writing, you need to follow the conventions of spelling and grammar (even if you decide to break those rules for special effect down the road).

Through the simple acts of capturing ideas, organizing them into groups, distilling the best parts, and assembling them together to create value for others, we are practicing the basic moves of knowledge work in such a way that we can improve on them over time.

This standardized routine is known as the *creative process*, and it operates according to timeless principles that can be found throughout history. By identifying the principles that stand the test of time despite huge changes in the underlying technology, we can better understand the essential nature of creativity.

The *products* of creativity are constantly changing and there is always a new "hot" trend to run after. One year it's Instagram photos, the next it's Snapchat stories, the next it's TikTok videos, and so on forever. Even the long tradition of the novel has evolved for each era.

But if you go one level deeper, to the *process* of creativity, it is a very different story. The creative process is ancient and unchanging. It was the same thousands of years ago as it is today. There are lessons we can learn on that deeper level that transcend any particular medium and any particular set of tools.

One of the most important patterns that underlies the creative process is called "divergence and convergence."[II]

Divergence and Convergence: A Creative Balancing Act

If you look at the process of creating anything, it follows the same simple pattern, alternating back and forth between divergence and convergence.

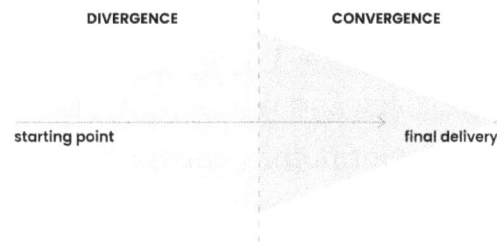

A creative endeavor begins with an act of divergence. You open the space of possibilities and consider as many options as possible. Like Taylor Swift's notes, Twyla Tharp's box, Francis Ford Coppola's prompt book, or Octavia Butler's commonplace notebooks, you begin to gather different kinds of outside inspiration, expose yourself to new influences, explore new paths, and talk to others about what you're thinking. The number of things you are looking at and considering is increasing—you are diverging from your starting point.

The activity of divergence is familiar to all of us: it is the classic whiteboard covered in sketches, the writer's wastepaper basket filled with crumpled-up drafts, and the photographer with hundreds of photos laid out across the floor. The purpose of divergence is to generate new ideas, so the process is necessarily spontaneous, chaotic, and messy. You can't fully plan or organize what you're doing in divergence mode, and you shouldn't try. This is the time to wander.

As powerful and necessary as divergence is, if all we ever do is diverge, then we never arrive anywhere. Like Francis Ford Coppola highlighting certain passages and crossing out others in *The Godfather* novel, at some point you must start discarding possibilities and converging toward a solution. Otherwise, you will never get the rewarding sense of completion that comes with hitting "send" or "publish" and stepping back from the canvas or screen knowing you got the job done.

Convergence forces us to eliminate options, make trade-offs, and decide what is truly essential. It is about narrowing the range of possibilities so that you can make forward progress and end up with a final result you are proud of. Convergence allows our work to take on a life of its own and become something separate from ourselves.

The model of divergence and convergence is so fundamental to all creative work, we can see it present in any creative field.

Writers diverge by collecting raw material for the story they want to tell, sketching out potential characters, and researching historical facts. They converge by making outlines, laying out plot points, and writing a first draft.

Engineers diverge by researching all the possible solutions, testing the boundaries of the problem, or tinkering with new tools. They converge by deciding on a particular approach, designing the implementation details, and bringing their blueprints to life.

Designers diverge by collecting samples and patterns, talking to users to understand their needs, or sketching possible solutions. They converge by deciding on a problem to solve, building wireframes, or translating their designs into graphics files.

Photographers diverge by taking photos of things they find interesting, juxtaposing different kinds of photos together, or experimenting with new lighting or framing techniques. They converge by choosing the shots for a collection, archiving unused images, and printing their favorites.

If we overlay the four steps of CODE onto the model of divergence and convergence, we arrive at a powerful template for the creative process in our time.

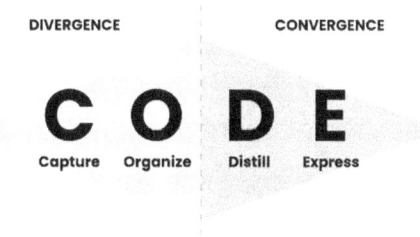

The first two steps of CODE, Capture and Organize, make up divergence. They are about gathering seeds of imagination carried on the wind and storing them in a secure place. This is where you research, explore, and add ideas. The final two steps, Distill and Express, are about convergence. They help us shut the door to new ideas and begin constructing something new out of the knowledge building blocks we've assembled.

The Three Strategies I Use to Bring Creative Work Together

Your Second Brain is a powerful ally in overcoming the universal challenge of creative work—sitting down to make progress and having no idea where to start.

Should you do more research, or start organizing the research you've already done?

Should you widen your horizons, or narrow your focus?

Should you start something new, or finish something you've already started?

When you distinguish between the two modes of divergence and convergence, you can decide each time you begin to work which mode you want to be in, which gives you the answers to the questions above. In divergence mode, you want to open up your horizons and explore every possible option. Open the windows and doors, click every link, jump from one source to another, and let your curiosity be your guide for what to do next. If you decide to enter convergence mode, do the opposite: close the door, put on noise-canceling headphones, ignore every new input, and ferociously chase the sweet reward of completion. Trust that you have enough ideas and enough sources, and it's time to turn inward and sprint toward your goal.

Of the two stages of this process, convergence is where most people struggle.

The more imaginative and curious you are, the more diverse your interests, and the higher your standards and commitment to perfection, the more difficult you will likely find it to switch from divergence mode into convergence mode. It's painful to cut off options and choose one path over another. There is a kind of creative grief in watching an idea that you know is full of potential get axed from a script or a story. This is what makes creative work challenging.

When you sit down to finish something—whether it's an explanatory email, a new product design, a research report, or a fundraising strategy—it can be so tempting to do more research. It's so easy to open up dozens of browser tabs, order more books, or go off in completely new directions. Those actions are tempting because they *feel* like productivity. They feel like forward progress, when in fact they are divergent acts that postpone the moment of completion.

There are three powerful strategies for completing creative projects I recommend to help you through the pitfalls of convergence. Each of them depends on having a Second Brain where you can manipulate and shape information without worrying it will disappear. Think of them as the tools in your Second Brain tool belt, which you can turn to anytime you need to get unstuck, find your way around obstacles, or decide what to do next.

1. The Archipelago of Ideas: Give Yourself Stepping-Stones

The Archipelago of Ideas technique is valuable any time you are starting a new piece of work—whether it's a how-to guide, a training workshop, a brief for a new project, or an essay you're publishing on your blog. It gives you a way to plan your progress even when performing tasks that are inherently unpredictable. The technique is named after a quote by Steven Johnson, the author of a series of fascinating books on creativity, innovation, and the history of ideas.[1] As Johnson wrote:

> *Instead of confronting a terrifying blank page, I'm looking at a document filled with quotes: from letters, from primary sources, from scholarly papers, sometimes even my own notes. It's a great technique for warding off the siren song of procrastination. Before I hit on this approach, I used to lose weeks stalling before each new chapter, because it was just a big empty sea of nothingness. Now each chapter starts life as a kind of archipelago of inspiring quotes, which makes it seem far less daunting. All I have to do is build bridges between the islands.*

An archipelago is a chain of islands in the ocean, usually formed by volcanic activity over long spans of time. The Hawaiian Islands, for example, are an archipelago of eight major islands extending over about 1,500 miles of the Pacific Ocean.

To create an Archipelago of *Ideas*, you divergently gather a group of ideas, sources, or points that will form the backbone of your essay, presentation, or deliverable. Once you have a critical mass of ideas to work with, you switch decisively into convergence mode and link them together in an order that makes sense.

Here's an example of an Archipelago of Ideas note I created to help me write an in-depth article on commonplace books:

> **Commonplace Books**
>
> ### Commonplace Books
>
> <u>7 Highlights from The Glass Box And The Commonplace Book</u>
>
> - In its most customary form, **"commonplacing," as it was called, involved transcribing interesting or inspirational passages from one's reading, assembling a personalized encyclopedia of quotations.**
> - The **philosopher John Locke first began maintaining a commonplace book in 1652, during his first year at Oxford**
> - The beauty of Locke's scheme was that it provided **just enough order to find snippets when you were looking for them, but at the same time it allowed the main body of the commonplace book to have its own unruly, unplanned meanderings.**
>
> <u>Write Quickly by Mining Your Commonplace Book</u>
>
> - **commonplace book: a centralized, personally curated, and continuously maintained collection of information**
> - Well, consider the painter Chuck Close, who works by deconstructing his huge images into small grids that he completes one at a time: I push little pieces of paint up against each other. **And I work essentially from the top down, left to right. And I slowly build these paintings—construct them the way that somebody might make a quilt or crochet or knit.**

The underlined links (which appear in green in my notes) are the sources I'm drawing on as research. Clicking a link will lead me not to the public web, where I can easily get distracted, but to another note *within* my Second Brain containing my full notes on that source.[III] There I will find all the details I might need, as well as a link back to the original work for my citations.

Below each source, I've copied and pasted only the points I specifically want to use in this particular piece of writing. This Archipelago of Ideas includes external sources as in my example above, but also notes I've taken based on my own thoughts and experiences. This gives me the best of both worlds: I can focus only on the relevant points right in front of me, but all the other details I might need are just a click away. The bolds and highlights of Progressive Summarization help me quickly determine which parts are most interesting and important at a glance.

The Archipelago of Ideas technique is a contemporary reinvention of the age-old practice of outlining—laying out the points you want to include up front, so that when it comes time to execute all you have to do is string them together. The note you see above is exactly what I will want in front of me when I sit down in convergence mode to finish the first draft of my article.

Creating outlines digitally instead of on paper offers multiple major advantages:

- **A digital outline is far more malleable and flexible**—you can add bullet points and cross them off, rearrange and expand on them, add bolding and highlights, and edit them after the fact as your thinking changes.

- **The outline can link to more detailed content**—instead of trying to cram every last point onto the same page, you can link to both your own private notes and public resources on the web, which helps avoid overloading the outline with too much detail.

- **The outline is interactive and multimedia**—you can add not only text, but images, GIFs, videos, attachments, diagrams, checkboxes, and more.

- **The outline is searchable**—even if it gets long, you have a powerful search feature to instantly call up any term you're looking for.

- **The outline can be accessed and edited from anywhere**—unlike a piece of paper in a notebook, your outline is instantaneously synced to every one of your devices, and it can be viewed, edited, and added to from anywhere.

An Archipelago of Ideas separates the two activities your brain has the most difficulty performing at the same time: *choosing* ideas (known as selection) and *arranging* them into a logical flow (known as sequencing).

The reason it is so difficult to perform these activities simultaneously is they require different modes: selection is divergent, requiring an open state of mind that is willing to consider any possibility. Sequencing is convergent, requiring a more closed state of mind focused only on the material you already have in front of you.

The goal of an archipelago is that instead of sitting down to a blank page or screen and stressing out about where to begin, you start with a series of small stepping-stones to guide your efforts. First you select the points and ideas you want to include in your outline, and then in a separate step, you rearrange and sequence them into an order that flows logically. This makes both of those steps far more efficient, less taxing, and less vulnerable to interruption.

Instead of starting with scarcity, start with abundance—the abundance of interesting insights you've collected in your Second Brain.

2. The Hemingway Bridge: Use Yesterday's Momentum Today

Ernest Hemingway was one of the most recognized and influential novelists of the twentieth century. He wrote in an economical, understated style that profoundly influenced a generation of writers and led to his winning the Nobel Prize in Literature in 1954.

Besides his prolific works, Hemingway was known for a particular writing strategy, which I call the "Hemingway Bridge." He would always end a writing session only when he knew what came next in the story. Instead of exhausting every last idea and bit of energy, he would stop when the next plot point became clear. This meant that the next time he sat down to work on his story, he knew exactly where to start. He built himself a bridge to the next day, using today's energy and momentum to fuel tomorrow's writing.[IV]

You can think of a Hemingway Bridge as a bridge between the islands in your Archipelago of Ideas. You may have the islands, but that is just the first step. The much more challenging work is linking them together into something that makes sense, whether it is a piece of writing, the design of an event, or a business pitch. The Hemingway Bridge is a way of making each creative leap from one island to the next less dramatic and risky: you keep some energy and imagination in reserve and use it as a launchpad for the next step in your progress.

How do you create a Hemingway Bridge? Instead of burning through every last ounce of energy at the end of a work session, reserve the last few minutes to write down some of the following kinds of things in your digital notes:

- **Write down ideas for next steps:** At the end of a work session, write down what you think the next steps could be for the next one.

- **Write down the current status:** This could include your current biggest challenge, most important open question, or future roadblocks you expect.

- **Write down any details you have in mind that are likely to be forgotten once you step away:** Such as details about the characters in your story, the pitfalls of the event you're planning, or the subtle considerations of the product you're designing.

- **Write out your intention for the next work session:** Set an intention for what you plan on tackling next, the problem you intend to solve, or a certain milestone you want to reach.

The next time you resume this endeavor, whether that's the next day or months later, you'll have a rich set of jumping-off points and next steps waiting for you. I often find that my subconscious mind keeps working in the background to help me improve on those thoughts. When I return to the project, I can combine the results of my past thinking with the power of a good night's sleep and put them together into a creative breakthrough.

To take this strategy a step further, there is one more thing you can do as you wrap up the day's work: send off your draft or beta or proposal for feedback. Share this Intermediate Packet with a friend, family member, colleague, or collaborator; tell them that it's still a work-in-process and ask them to send you their thoughts on it. The next time you sit down to work on it again, you'll have their input and suggestions to add to the mix of material you're working with.

3. Dial Down the Scope: Ship Something Small and Concrete

A third technique I recommend for convergence I call "Dial Down the Scope."

"Scope" is a term from project management that has been adopted by software developers, from whom I learned it while working in Silicon Valley. The scope refers to the full set of features a software program might include.

Let's say you're designing a fitness app. You sketch out a beautiful vision: it will have workout tracking, calorie counts, a gym finder, progress charts, and even connect you with others via a social network. It's going to be amazing! It will transform people's lives!

As with so many ambitious goals, once you get into the details it dawns on you just how complex these features are to build. You have to design the user interface, but also build the backend system to make it work. You have to hire customer support representatives and train them how to troubleshoot problems. You need a whole finance operation to keep track of payments and comply with regulations. Not to mention all the responsibilities of managing employees, dealing with investors, and developing a long-term strategy.

The solution that software teams landed on to deal with this kind of ballooning complexity is to "dial down the scope." Instead of postponing the release of the app, which might prove disastrous in the face of looming competition and only delays the learning they need, the development team starts "dialing down" features as the release date approaches. The social network gets postponed to a future version. The progress charts lose their interactive features. The gym finder gets canceled completely. The first parts to be dialed down are the ones that are most difficult or expensive to build, that have the most uncertainty or risk, or that aren't central to the purpose of the app. Like a hot-air balloon trying to take off, more and more features get thrown overboard to lighten the load and get the product off the ground. Any features that don't make it into this version can always be released as part of future software updates.

How does this relate to our careers as knowledge workers?

We also deliver complex pieces of work under strict deadlines. We also have limited time, money, attention, and support—there are always constraints we must work within.

When the full complexity of a project starts to reveal itself, most people choose to delay it. This is true of projects at work, and even more true of side projects we take on in our spare time. We tell ourselves we just need more time, but the delay ends up creating more problems than it solves. We start to lose motivation as the time horizon stretches out longer and longer. Things get lost or go out of date. Collaborators move on, technology becomes obsolete and needs to be upgraded, and random life events never fail to interfere. Postponing our goals and desires to "later" often ends up depriving us of the very experiences we need to grow.

The problem isn't a lack of time. It is that we forget that we have control over the *scope* of the project. We can "dial it down" to a more manageable size, and we must if we ever want to see it finished.

Waiting until you have everything ready before getting started is like sitting in your car and waiting to leave your driveway until all the traffic lights across town are green at the same time. You can't wait until everything is perfect. There will always be something missing, or something else you think you need. Dialing Down the Scope recognizes that not all the parts of a given project are equally important. By dropping or reducing or postponing the least important parts, we can unblock ourselves and move forward even when time is scarce.

Your Second Brain is a crucial part of this strategy, because you need a place to save the parts that get postponed or removed.

You might cut sentences or entire pages from an article you're writing, or delete scenes from a video you're making, or drop parts of a speech when you're trying to keep within your allotted time. This is a completely normal and necessary part of any creative process.

That doesn't mean you have to throw away those parts. One of the best uses for a Second Brain is to collect and save the scraps on the cutting-room floor in case they can be used elsewhere. A slide cut from a presentation could become a social media post. An observation cut from a report could become the basis for a conference presentation. An agenda item cut from a meeting could become the starting point for the next meeting. You never know when the rejected scraps from one project might become the perfect missing piece in another. The possibilities are endless.

Knowing that nothing I write or create truly gets lost—only saved for later use—gives me the confidence to aggressively cut my creative works down to size without fearing that I've wasted effort or that I'll lose the results of my thinking forever. Knowing that I can always release a fix, update, or follow up on anything I've made in the past gives me the courage to share my ideas before they're perfectly ready and before I have them all figured out. And sharing before I feel ready has completely altered the trajectory of my career.

Whatever you are building, there is a smaller, simpler version of it that would deliver much of the value in a fraction of the time. Here are some examples:

- If you want to write a book, you could dial down the scope and write a series of online articles outlining your main ideas. If you don't have time for that, you could dial it down even further and start with a social media post explaining the essence of your message.

- If you want to deliver a workshop for paying clients, you could dial it down to a free workshop at a local meetup, or dial it down even further and start with a group exercise or book club for a handful of colleagues or friends.

- If you'd like to make a short film, start with a YouTube video, or if that's too intimidating, a livestream. If it's still too much, record a rough cut on your phone and send it to a friend.

- If you want to design a brand identity for a company, start with a mock-up of a single web page. Even easier, start with a few hand-drawn sketches with your ideas for a logo.

How can you know which direction to take your thinking without feedback from customers, colleagues, collaborators, or friends? And how can you collect that feedback without showing them something concrete? This is the chicken-and-egg problem of creativity: you don't know what you should create, but you can't discover what people want until you create something. Dialing Down the Scope is a way of short-circuiting that paradox and testing the waters with something small and concrete, while still protecting the fragile and tentative edges of your work.

Divergence and convergence are not a linear path, but a loop: once you complete one round of convergence, you can take what you've learned right back into a new cycle of divergence. Keep alternating back and forth, making iterations each time until it's something you can consider "done" or "complete" and share more widely.

Divergence and Convergence in the Wild: Behind the Scenes of a Home Project

Let me share an example of one of my own projects in which I used all three of these techniques: remodeling our garage into a home office.

When we moved into our home, my wife and I soon realized we needed a better workspace. We both work from home, and the tiny extra bedroom wasn't cutting it, especially once our son was born. We excitedly made plans to turn our garage into a home studio. The moment I created a dedicated project folder, I knew it was on.

I started by creating an Archipelago of Ideas—an outline of the main questions, considerations, desired features, and constraints I thought our project would entail. Here is the outline I came up with after fifteen minutes:

> **Project brief: Forte Academy Studio**
>
> ## Project brief: Forte Academy Studio
>
> Intro
> - **Extremely multi-functional, modular, and flexible according to changing needs**
> - **Also use as meeting space/home office**
>   - How to Work Productively From Home Without Going Crazy: Tiago's Top 10 WFH Tips
> - **Bathroom/living space ADU**
>   - **Can we have mini-kitchen?**
>
> Cost
>
> Ideas
> - **Virtual interactive experiences (VIEs)**
> - **Straddling in-person and digital worlds**
> - My two biggest inspirations are this video of Tony Robbins' setup for UPW, and this video of the cyberillusionist Marco Tempest doing a keynote address from his home studio
> - Tweetstorm on modern learning
>
> Phases/timeline
> - **Phase 1: Garage remodel/home office**
> - **Phase 2: Broadcasting studio**
> - **Phase 3: Recording studio**
>
> Zoom setup needs/backdrops
> - **Deep background for creating depth behind the subject**
> - Equipment

I didn't know up front what the main headings of this document would include, but as I wrote out my thoughts they soon emerged: Intro, Cost, Ideas, Phases, Aesthetics, Zoom Setup, Equipment, and Open Questions. I did a few searches of my Second Brain for terms like "home office" and "home studio" and found several existing notes that could come in handy as well. For example, I found the notes with recommendations from a friend who had experience designing studios, which I mentioned previously; photos of a beautifully designed café in Mexico City that my wife and I loved visiting and wanted to mimic; and a note with best practices for hosting Zoom calls, such as finding the right lighting and a background that isn't too distracting. I added links to them at the bottom of my outline as well.

Even with some existing material to work with, there were gaps in our plan. Over the next few weeks, whenever I had a free window of time, I collected and captured tidbits of content to inform our home studio remodel. I saved photos from Pinterest showing examples of home offices that I thought looked neat; notes from a conversation with a musician friend who taught me about soundproofing; and a list of local contractors a neighbor shared with me to reach out to. I even went on a late-night spree watching dozens of videos of YouTubers giving tours of their studios, taking notes on the small details of how they converted empty spaces into functional workspaces.

Between managing the business and our household, my time was incredibly scarce as we began the remodel. Whenever I could, I would highlight and distill the last few notes I'd captured and leave my future self a brief note about where I left off. I used a series of Hemingway Bridges to string together many such windows of time that otherwise wouldn't have been of much productive use.

Finally, as all these thoughts and ideas and wishes and dreams began to add up, the project became quite a juggernaut. Before I knew it, our ambitions had expanded to knocking down walls, cutting through the roof for a skylight, laying down cable for superfast Internet, and redesigning the layout of the backyard to accommodate it all. We had diverged too far and needed to rein it in a bit.

This is where Dialing Down the Scope was essential: we identified the most outlandish of our plans and decided to save those for a later stage. I moved those ideas to their own "someday/maybe" section of my outline to revisit later. My wife and I also added several constraints to the project, such as the budget we were willing to spend, and a deadline to have the remodel done by a certain date. These constraints helped us reduce the scope of the project to something reasonable and manageable. As soon as we did, the next steps of finding a contractor and finalizing the floorplan became crystal clear.

Your Turn: Move Fast and Make Things

If you'd like to give this approach to executing projects a try, now is the perfect time.

Start by picking one project you want to move forward on. It could be one you identified in Chapter 5, when I asked you to make folders for each active project. It could alternatively be something you know you want to (or have to) get started on. The more uncertain, new, or challenging the project, the better.

Make an outline with your goals, intentions, questions, and considerations for the project. Start by writing out anything already on your mind, and then peruse your PARA categories for related notes and Intermediate Packets. These could include points or takeaways from previously created notes, inspiration from models or examples you want to borrow from, or templates you can use to follow best practices.

Here are some useful questions to ask as you conduct your search:

- Is there a book or article you could extract some excerpts from as inspiration?

- Are there websites that might have resources you could build upon?

- Are there podcasts by experts you could subscribe to and listen to while commuting or doing household chores?

- Are there relevant IPs buried in other projects you've worked on in the past?

Some material you find will be very succinct and highly polished, while some might be quite rough. It doesn't matter—your only goal is to get all the potentially usable material in one place. Move all the notes and IPs you might want to use into a new project folder.

Set a timer for a fixed period of time, such as fifteen or twenty minutes, and in one sitting see if you can complete a first pass on your project *using only the notes you've gathered in front of you*. No searching online, no browsing social media, and no opening multiple browser tabs that you swear you're going to get to eventually. Only work with what you already have. This first pass could be a plan, an agenda, a proposal, a diagram, or some other format that turns your ideas into a tangible artifact.

You might experience some FOMO—that inner Fear of Missing Out—that pushes you to seek out yet another morsel of information somewhere out there. You will probably be tempted to go off and "do more research," but you are not *completing* the entire project in one sitting. You are only creating the first iteration—a draft of your essay, a sketch of your app, a plan for your campaign. Ask yourself, "What is the smallest version of this I can produce to get useful feedback from others?"

If you find that you can't complete the first iteration in one sitting, start by building a Hemingway Bridge to the next time you can work on it. List open questions, remaining to-dos, new avenues to explore, or people to consult. Share what you've produced with someone who can give you feedback while you're away and save their comments in a new note in the same project folder. You can collect this feedback in a private conversation with a trusted colleague, or publicly on social media at full blast, or anywhere in between. Pick a venue for sharing that you feel comfortable with.

If you feel resistance to continuing with this project later, try Dialing Down the Scope. Drop the least important features, postpone the hardest decisions for later, or find someone to help you with the parts you're least familiar with.

Throughout every step of this process, be sure to keep notes on anything you learn or discover, or any new Intermediate Packets you might want to seek out. Once your biological brain is primed by this first pass through your notes, you'll start to notice signs and clues related to it everywhere you look. Save those clues as notes as well! Once you're finished with your first iteration, have gathered feedback, and collected a new set of notes to work with, you'll be ready for whatever comes next.

I. For more insight into what I learned about the creative process from my father, I made a short documentary on his work and life called *Wayne Lacson Forte: On My Way To Me*.

II. I first learned about the model of divergence and convergence from Design Thinking, an approach to creative problem-solving that emerged out of the Stanford Design School and was further popularized by the innovation consultancy IDEO starting in the 1980s and 1990s.

III. If your notes app syncs with your computer, this also means that you can disconnect from the Internet and still make progress, since you have all your notes saved on your hard drive.

IV. One way to think of this is to "end with the beginning in mind," a clever rephrasing of author Stephen Covey's classic advice to "begin with the end in mind."

Chapter 9

The Essential Habits of Digital Organizers

> Habits reduce cognitive load and free up mental capacity, so you can allocate your attention to other tasks… It's only by making the fundamentals of life easier that you can create the mental space needed for free thinking and creativity.
> —James Clear, author of *Atomic Habits*

Your Second Brain is a practical system for enhancing your productivity *and* your creativity.

While these domains are often seen as mutually exclusive opposites—one concrete and defined; the other abstract and open-ended—instead I see them as complementary. When we are organized and efficient, that creates space for creativity to arise. When we have confidence in our creative process, we don't have to think about it as much, significantly reducing the background stress of constantly worrying whether we're going in the right direction.

This balance between order and creativity is something that we can build into our Second Brain intentionally. Like every system, a Second Brain needs regular maintenance. There is a certain level of organization that you want to maintain in your digital world, so that when you go there to get things done, your virtual workspaces support your productivity instead of interfering with it.

"Being organized" isn't a personality trait you're born with, nor is it merely a matter of finding the right apps or tools. Being organized is a habit—a repeated set of actions you take as you encounter, work with, and put information to use. If we're constantly scrambling to find our notes, drafts, brainstorms, and sources, not only do we waste precious time, but we also sabotage our momentum. At each step of CODE, there are habits that can help us be more organized so that our creativity has space to arise.

The Mise-en-Place Way to Sustainable Productivity

Consider how chefs work in a commercial kitchen. They have incredibly high demands on both the quality *and* quantity of their output. Every ingredient in every dish must be nearly perfect—one cold side or undercooked filet and the whole dish can be sent back, and the kitchen might have to produce hundreds of dishes on a busy night.

This fundamental tension—between quality and quantity—is a tension we share as knowledge workers. We also must produce work to an extremely high standard, and we must do it fast, continuously, all year long. We are like sprinters who are also trying to run a marathon.

Chefs have a particular system for accomplishing this daunting feat. It's called *mise en place*, a culinary philosophy used in restaurants around the world. Developed in France starting in the late 1800s, mise en place is a step-by-step process for producing high-quality food efficiently. Chefs can never afford to stop the whole kitchen just so they can clean up. They learn to keep their workspace clean and organized *in the flow of the meals they are preparing*.

In the kitchen, this means small habits like always putting the mixing spoon in the same place so they know where to find it next time; immediately wiping a knife clean after using it so it's ready for the next cut; or laying out the ingredients in the order they'll be used so that they serve as placeholders.

Chefs use mise en place—a philosophy and mindset embodied in a set of practical techniques—as their "external brain."[1] It gives them a way to externalize their thinking into their environment and automate the repetitive parts of cooking so they can focus completely on the creative parts.

We have a lot to learn as knowledge workers from the system of mise en place. We likewise have to contend with a deluge of tasks, under uncertain conditions, with tight deadlines. We also receive a constant stream of inputs and requests, have too little time to process them, and face many demands requiring simultaneous attention. For us as well, the only time we have available to maintain our systems is during the execution of our regular work.

There's no time that's magically going to become available for you to stop everything and completely reorganize your digital world. It's not likely that your manager is going to look kindly on you blocking off a whole day to "get caught up." Your business won't last long if you turn customers away because you're "maintaining your systems." It's difficult to find the time to put the world on hold and catch your breath. We tend to notice our systems need maintenance only when they break down, which we then blame on our lack of self-discipline or our failure to be sufficiently productive.

Building a Second Brain is not just about downloading a new piece of software to get organized at one point in time; it is about adopting a dynamic, flexible system and set of habits to continually access what we need without throwing our environment (and mind) into chaos.

It's not enough to have inner discipline. We also need to follow an *outer* discipline—a system of principles and behaviors—to channel our energies, thoughts, and emotions productively. A system that adds some structure to the constantly changing flux of information that we interact with every day.

In this chapter, I will introduce you to three kinds of habits that can be integrated into your routine to ensure your Second Brain remains functional and relevant. Each of these habits creates boundaries—of time, space, and intention—around the states of mind that you want to protect and promote in your life. These boundaries tell you what you should be focusing on, and just as importantly, what you should ignore. The three habits most important to your Second Brain include:

- **Project Checklists:** Ensure you start and finish your projects in a consistent way, making use of past work.

- **Weekly and Monthly Reviews:** Periodically review your work and life and decide if you want to change anything.

- **Noticing Habits:** Notice small opportunities to edit, highlight, or move notes to make them more discoverable for your future self.

You can think of these habits as the "maintenance schedule" of your Second Brain. Just like you have a maintenance schedule for your car, which advises you to regularly change the oil, rotate the tires, and change the air filters, your Second Brain occasionally needs a tune-up to ensure it's in good working order.

Let's explore these habits one at a time.

The Project Checklist Habit: The Key to Starting Your Knowledge Flywheel

At the most basic level, knowledge work is about taking in information and then turning it into results. All day, every day we are consuming and then producing. You don't need special training to perform these activities, and you certainly don't need a Second Brain.

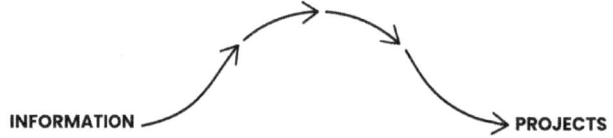

What most people are missing, however, is a feedback loop—a way to "recycle" the knowledge that was created as part of past efforts so it can be used in future ones as well. This is how investors think about money: they don't get the profits from one investment and immediately spend it all. They reinvest it back into other investments, creating a flywheel so their money builds on itself over time.

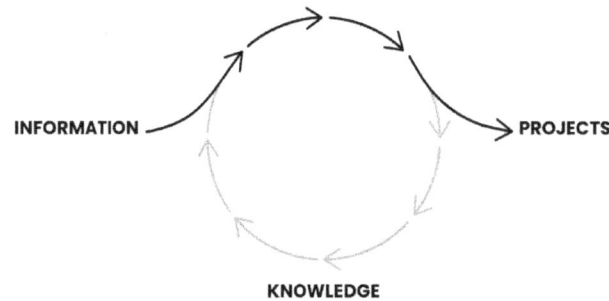

This is exactly how I want you to treat your attention—as an asset that gets invested and produces a return, which in turn can be reinvested back into other ventures. This is how you can ensure your knowledge grows and compounds over time like a high-yield asset. Like investing a small amount in the stock market every month, your investments of attention can likewise compound as your knowledge grows and your ideas connect and build on each other.

If you look closely, there are two key moments in this process of recycling knowledge. Two places where the paths diverge, and you have the chance to do something different than you've done before.

Those two moments are when a project starts, and when it finishes. For the former, I'll introduce you to the Project Kickoff Checklist, and for the latter, the Project Completion Checklist.

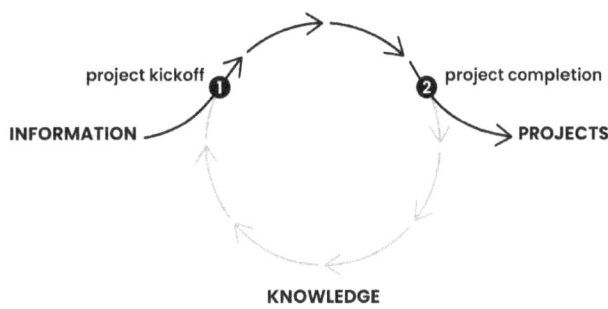

Checklist #1: Project Kickoff

Before they taxi onto the runway and take off, airline pilots run through a "preflight checklist" that tells them everything they need to check or do. It ensures they complete all the necessary steps without having to rely on their unreliable brains.

The way most people launch projects, in contrast, can be described as "haphazardly." They might look through their existing notes and files for any information that might be relevant, or they might not. They might talk to their colleagues about any lessons from past experience, or they might not. They might create a plan to guide their progress, or they might not. The successful start of the project is more or less left to chance.

In Chapter 5 we saw how work is becoming ever more project-centric. Every goal, collaboration, or assignment we take on can be defined as a project, which gives it shape, focus, and a sense of direction. If we consider that these projects are our biggest investments of attention, it's worth adding a little bit of structure to how we start them. This is where the Project Kickoff Checklist comes in.

Here's my own checklist:

1. **Capture** my current thinking on the project.
2. **Review** folders (or tags) that might contain relevant notes.
3. **Search** for related terms across all folders.
4. **Move** (or tag) relevant notes to the project folder.
5. **Create** an outline of collected notes and plan the project.

1. Capture my current thinking on the project. I often find that the moment a project begins to form in my mind, I start to have ideas and opinions about it. I like to start by creating a blank note and doing a brainstorm of any thoughts that come to mind. This first note is then placed inside a new project folder dedicated to storing all the notes I'll be creating related to it.

This step can and should be messy: I pour out all my random musings, potential approaches, links to other ideas or topics, or reminders of people to talk to.

Here are some questions I use to prompt this initial brainstorm:

- What do I already know about this project?
- What don't I know that I need to find out?
- What is my goal or intention?
- Who can I talk to who might provide insights?
- What can I read or listen to for relevant ideas?

Anything that comes to mind from these questions I write down in my starting note. I prefer using bullet points so the information is compact and can easily be moved around.

2. **Review folders (or tags) that might contain relevant notes.** Second, I look through any existing folders that might contain information relevant to the new project I'm starting, including related templates, outlines, and outtakes from previous projects. PARA and Progressive Summarization really come in handy here: I already have a variety of folders, each containing a curated set of notes, highlighted and summarized so I can rapidly recall what they're about. I choose a handful of folders that seem most relevant to what I'm starting, including in projects, areas, resources, and archives. Then I quickly scan any notes inside that look interesting, taking care to keep the momentum going so I don't get bogged down anywhere. Now is not the time to go on tangents that will only distract me from moving forward.

3. **Search for related terms across all folders.** The third step is to perform searches for any notes I might have missed. Sometimes there are valuable ideas buried in unexpected places, which I may not find through browsing alone.

This is where the Curator's Perspective I used when I first captured the content really pays off—because each and every note in my Second Brain was deliberately chosen, I am able to search through a collection of exclusively high-quality notes free of fluff and filler. This is in stark contrast to searching the open Internet, which is full of distracting ads, misleading headlines, superficial content, and pointless controversy, all of which can throw me off track.

I run a series of searches for terms related to the new project, scanning the results and quickly jumping into any note that seems relevant. Progressive Summarization helps here too, enabling me to zoom into and out of notes without having to absorb their full contents.

4. **Move (or tag) relevant notes to the project folder.** Fourth, any notes found in the previous two steps I move to the project folder, titled after the name of the new project I'm starting. Alternatively, depending on the capabilities of your

notes app, you can also tag or link any relevant notes with the project, so you don't have to move them from their original location. The important thing isn't where a note is located, but whether you can reference it quickly while staying focused on the project at hand.

5. **Create an outline of collected notes and plan the project.** Finally, it's time to pull together the material I've gathered and create an outline (an Archipelago of Ideas) for the project. My goal is to end up not just with a loose collection of ideas. It is to formulate a logical progression of steps that make it clear what I should do next.

The form this outline takes depends on the nature of the project. If it is a piece of writing such as an essay or report, the outline might be the main points or headings I want to include in the final piece. If it is a document outlining a collaborative project with colleagues or outside contractors, the outline might include the objectives we're working toward and tentative responsibilities for each person. If it is a trip I'm planning to take, it might be a packing list and itinerary.

The important thing to remember as you move through this checklist is that you are making a plan for how to tackle the project, *not executing the project itself*. You should think of this five-step checklist as a first pass, taking no more than twenty to thirty minutes. You're only trying to get a sense of what kind of material you already have in your Second Brain. Once you do, you'll have a much better sense of how much time it will take, which knowledge or resources you'll need access to, and what your challenges will likely be.

I encourage you to use my kickoff checklist as a starting point and customize it over time as you understand how it fits into your own context. Depending on your profession or industry, you might need more or less formality, more or less time for a first pass, and more or fewer people involved. Here are some other options for actions you might want to include in your own version:

- **Answer premortem[1] questions:** What do you want to learn? What is the greatest source of uncertainty or most important question you want to answer? What is most likely to fail?

- **Communicate with stakeholders:** Explain to your manager, colleagues, clients, customers, shareholders, contractors, etc., what the project is about and why it matters.
- **Define success criteria:** What needs to happen for this project to be considered successful? What are the minimum results you need to achieve, or the "stretch goals" you're striving for?
- **Have an official kickoff:** Schedule check-in calls, make a budget and timeline, and write out the goals and objectives to make sure everyone is informed, aligned, and clear on what is expected of them. I find that doing an official kickoff is useful even if it's a solo project!

Checklist #2: Project Completion

Now let's take a look at the Project Completion Checklist, the other side of the equation.

The completion of a project is a very special time in a knowledge worker's life because it's one of the rare moments when something actually ends. Part of what makes modern work so challenging is that nothing ever seems to finish. It's exhausting, isn't it? Calls and meetings seem to stretch on forever, which means we rarely get to celebrate a clear-cut victory and start fresh. This is one of the best reasons to keep our projects small: so that we get to feel a fulfilling sense of completion as often as possible.

We don't want to limit ourselves to merely celebrating the end of a project. We want to learn from the experience and document any thinking that could add value in the future. This is where the Project Completion Checklist is essential. It's a series of steps you can take to decide if there are any reusable knowledge assets worth keeping, before archiving the rest. The only way that the Kickoff Checklist we just looked at will be feasible is if you've previously taken the time to save and preserve material from past projects.

Here's my checklist:

1. **Mark** project as complete in task manager or project management app.

2. **Cross out** the associated project goal and move to "Completed" section.

3. **Review** Intermediate Packets and move them to other folders.

4. **Move** project to archives across all platforms.

5. **If project is becoming inactive:** add a current status note to the project folder before archiving.

1. **Mark project as complete in task manager or project management app.** This is the first step is making sure the project is in fact finished. Often there are a few lingering tasks needed to completely wrap it up—such as getting final approvals, filing paperwork, or disseminating the project deliverables—which is why I start by looking at my task manager. A task manager is a dedicated app for keeping track of pending actions, like a digital to-do list.[II]

If all the tasks I find there are done, I can mark it as complete and move on to the following steps.

2. **Cross out the associated project goal and move to "Completed" section.** Each project I work on usually has a corresponding goal. I keep all my goals in a single digital note, sorted from short-term goals for the next year to long-term goals for years to come.

I like to take a moment and reflect on whether the goal I initially set for this project panned out. If I successfully achieved it, what factors led to that success? How can I repeat or double down on those strengths? If I fell short, what happened? What can I learn or change to avoid making the same mistakes next time? The amount of time you spend thinking about these questions depends on the size of the project. A massive team endeavor might justify hours of in-depth analysis, while a small personal side project might deserve only a few minutes of reflection.

I also like to cross out the goal and move it to a different section called "Completed." Any time I need some motivation, I can look through this list and be reminded of all the meaningful goals I've achieved in the past. It doesn't matter if the goal is big or small—keeping an inventory of your victories and successes is a wonderful use for your Second Brain.

3. **Review Intermediate Packets and move them to other folders.** Third, I'll look through the folder for the completed project to identify any Intermediate Packets I created that could be repurposed in the future. This could include a web-page design to be used as a template for future sites, an agenda for a one-on-one performance review, or a series of interview questions that might come in handy for future hires.

It takes a certain lens to see each of these documents and files not as disposable, but as tangible by-products of quality thinking. Much of our work gets repeated over time with slight variations. If you can start your thinking where you left off last time, you'll be far ahead compared to starting from zero every time.

Any IPs I decide could be relevant to another project, I move to that project's folder. The same goes for notes relevant to areas or resources. This is a forgiving decision, and it's okay if you don't catch every single one. The full contents of everything you archive away will always show up in future searches, so you don't have to worry that anything will be lost.

4. **Move project to archives across all platforms.** Fourth, it's time to move the project folder to the archives in my notes app, as well as any other platforms I used during the project. For me, this usually includes my computer's documents folder and my cloud storage drive.

This move ensures that your list of active projects doesn't get cluttered with old, obsolete stuff from the past while also preserving every bit of material just in case it unexpectedly becomes relevant in the future.

5. **If project is becoming inactive: add a current status note to the project folder before archiving.** The fifth step applies only if the project is getting canceled, postponed, or put on hold instead of completed. I still want to archive it so it's out of sight, but in this special case, there's one final action I take.

I add a new note to the project folder titled "Current status," and jot down a few comments so I can pick it back up in the future. For example, in a few bullet points I might describe the last actions I took, details on why it was postponed or canceled, who was working on it and what role they played, and any lessons or best practices learned. This Hemingway Bridge gives me the confidence to put the project on ice knowing I can bring it back to life anytime.

I've been amazed that by being honest with myself about when a project has stalled and taking these few minutes to download my current thinking on it, I can often pick it back up months or even years later with minimal effort. It's very empowering to realize you can put a project in "cold storage" and let go of the mental toll of having to keep it in mind. It's tremendously comforting to know that I don't need to make constant progress on everything all the time.

Here are some other items you can include on your Project Completion Checklist. I encourage you to personalize it for your own needs:

- **Answer postmortem questions:** What did you learn? What did you do well? What could you have done better? What can you improve for next time?

- **Communicate with stakeholders:** Notify your manager, colleagues, clients, customers, shareholders, contractors, etc., that the project is complete and what the outcomes were.

- **Evaluate success criteria:** Were the objectives of the project achieved? Why or why not? What was the return on investment?

- **Officially close out the project and celebrate:** Send any last emails, invoices, receipts, feedback forms, or documents, and celebrate your accomplishments with your team or collaborators so you receive the feeling of fulfillment for all the effort you put in.

The first pass on your Project Completion Checklist should be completed in even less time than the Project Kickoff Checklist—no more than ten or fifteen minutes to grab any stand-alone materials and insights. Since you don't know for sure that any of this material will ever be useful again, you should minimize how much extra time and attention you invest in it. Put in just enough effort that your future self will be able to decide if the material is relevant to their needs. If it is, then they can decide in that moment whether to invest the effort to further organize and distill it.

The purpose of using project checklists isn't to make the way you work rigid and formulaic. It is to help you start and finish projects cleanly and decisively, so you don't have "orphaned" commitments that linger on with no end in sight. Think of these checklists as scaffolding—a supporting structure that ensures what you're building can stand on its own. Just as scaffolding eventually gets taken down, these habits will get absorbed into the way you think and become completely second nature. You won't even consider starting something new without querying your Second Brain to see if there is any material you can reuse.

The Review Habit: Why You Should Batch Process Your Notes (and How Often)

Next let's talk about Weekly and Monthly Reviews.

The practice of conducting a "Weekly Review" was pioneered by executive coach and author David Allen in his influential book *Getting Things Done*.[III] He described a Weekly Review as a regular check-in, performed once a week, in which you intentionally reset and review your work and life. Allen recommends using a Weekly Review to write down any new to-dos, review your active projects, and decide on priorities for the upcoming week.

I suggest adding one more step: review the notes you've created over the past week, give them succinct titles that tell you what's inside, and sort them into the appropriate PARA folders. Most notes apps have an "inbox" of some kind where new notes collect until they're ready to be reviewed. This "batch processing" takes only seconds per note, and you can complete it within a few minutes.

Let's dive into the details and see how Weekly and Monthly Reviews can help you maintain your Second Brain in a state of readiness for whatever arrives at your doorstep.

A Weekly Review Template: Reset to Avoid Overwhelm

Here is my own Weekly Review Checklist, which I usually complete every three to seven days depending on how busy a given week is. The point isn't to follow a rigid schedule, but to make it a habit to empty my inboxes and clear my digital workspaces on a regular basis to keep them from getting overwhelmed. I keep this checklist on a digital sticky note on my computer, so I can easily refer to it.

1. Clear my email inbox.
2. Check my calendar.
3. Clear my computer desktop.
4. Clear my notes inbox.
5. Choose my tasks for the week.

1. **Clear my email inbox.** I start by clearing my email inbox of any emails lingering from the past week. I don't usually have time to do this during the week in the rush of my other priorities, but I've found that if I let messages accumulate from one week to the next, it makes it hard to figure out what's new and requires action and what's left over from the past.

Any action items I find get saved in my task manager, and any notes I capture get saved in my notes app.

2. **Check my calendar.** Next, I check my calendar. This is the landscape of my week, showing me the meetings and appointments I need to make room for. I typically look at the last couple of weeks in case there's anything I need to follow up on, and the upcoming couple of weeks in case there's anything I need to prepare for.

Once again, anything I need to act on gets saved in my task manager, and any notes get captured in my notes app.

3. **Clear my computer desktop.** Next, I clear the files that have accumulated on my computer desktop. I've found that if I let them accumulate week after week, eventually my digital environment gets so cluttered that I can't think straight.

Any files potentially relevant to my projects, areas, or resources get moved to the appropriate PARA folders in my computer's file system.

4. **Clear my notes inbox.** By the time I get to the fourth step, the inbox in my notes app is chock-full of interesting tidbits from the previous three steps—from my email, calendar, and computer desktop. Plus all the other notes I've collected over the course of the preceding week, which usually totals between five and fifteen new notes in an average week.

At this point I'll batch process them all at once, making quick, intuitive decisions about which of the PARA folders each note might be relevant to, and creating new folders as needed. There is no "correct" location for a given note, and search is incredibly effective, so I put it in the first place that occurs to me.

You'll notice that this is the only step in my Weekly Review that is directly related to my digital notes. It is a simple and practical process of going through my notes inbox, giving each note an informative title, and moving them into the appropriate PARA folders. I don't highlight or summarize them. I don't try to understand or absorb their contents. I don't worry about all the topics they could potentially relate to.

I want to save all that thinking for the future—for a time and place when I know what I'm trying to accomplish and am seeking a knowledge building block to help me get there faster. This weekly sorting process serves as a light reminder of the knowledge I've accumulated over the past week, and ensures I have a healthy flow of new ideas and insights flowing into my Second Brain.

5. **Choose my tasks for the week.** There's one final step in my Weekly Review. It's time to clear the inbox in my task manager app. By this point, there are likewise a number of tasks that I've captured from my email, calendar, desktop, and notes, and I take a few minutes to sort them into the appropriate projects and areas.

The final step of my Weekly Review is to select the tasks I'm committing to for the upcoming week. Because I've just completed a sweep of my entire digital world and taken into account every piece of potentially relevant information, I can make this decision decisively and begin my week with total confidence that I'm working on the right things.

A Monthly Review Template: Reflect for Clarity and Control

While the Weekly Review is grounded and practical, I recommend doing a Monthly Review that is a bit more reflective and holistic. It's a chance to evaluate the big picture and consider more fundamental changes to your goals, priorities, and systems that you might not have the chance to think about in the busyness of the day-to-day.

Here's mine:

1. Review and update my goals.
2. Review and update my project list.
3. Review my areas of responsibility.
4. Review someday/maybe tasks.
5. Reprioritize tasks.

1. **Review and update my goals.** I start by reviewing my goals for the quarter and the year. I ask myself questions like "What successes or accomplishments did I have?" and "What went unexpectedly and what can I learn from it?" I'll take some time to cross off any completed goals, add any new ones that have emerged, or change the scope of goals that no longer make sense.

2. **Review and update my project list.** Next, I'll review and update my project list. This includes archiving any completed or canceled projects, adding any new ones, or updating active projects to reflect how they've changed. I will also update the folders in my notes app to reflect these changes.

It's important that the project list remains a current, timely, and accurate reflection of your real-life goals and priorities. Especially since projects are the central organizing principle of your Second Brain. When you have a project folder ready and waiting, your mind is primed to notice and capture the best ideas to move it forward.

3. **Review my areas of responsibility.** Now it's time to do the same for my areas of responsibility. I'll think about the major areas of my life, such as my health, finances, relationships, and home life, and decide if there's anything I want to

change or take action on. This reflection often generates new action items (which go into my task manager) and new notes (which get captured in my notes app).

Area notebooks often contain notes that become the seeds of future projects. For example, I used an area folder called "Home" to collect photos for the home studio remodel I mentioned previously. Even before it was an active project, that broader area gave me a place to collect ideas and inspiration so it was ready and waiting the moment we decided to get started.

4. **Review someday/maybe tasks.** "Someday/maybe" is a special category for things I'd like to get to someday, but not in the near future. Things like "Learn Mandarin" and "Plant an orchard." These kinds of future dreams are important to keep track of, but you don't want them cluttering your priorities today. I'll take a few minutes to go through my "someday/maybe" tasks just in case any of them have become actionable. For example, when my wife and I settled down and became homeowners, our dream of getting a dog, which was impossible while moving from one apartment to another, suddenly came within reach. I had already saved a few notes on the kinds of dogs we should consider (athletic, hypoallergenic, good with kids, etc.), and this step in my Monthly Review reminded me to bring them to the surface.

5. **Reprioritize tasks.** Finally, once I've completed all the previous steps and have a holistic picture of my goals and projects in mind, it's time to reprioritize my tasks. I'm often surprised just how much can change in a month. To-dos that seemed critical last month might become irrelevant this month, and vice versa.

The Noticing Habits: Using Your Second Brain to Engineer Luck

There's a third category of habits that will come in handy as you start putting your Second Brain into action in the real world. It is in some ways the most important category, but also the least predictable.

I call them "noticing" habits—taking advantage of small opportunities you notice to capture something you might otherwise skip over or to make a note more actionable or discoverable. Here are some examples:

- Noticing that an idea you have in mind could potentially be valuable and capturing it instead of thinking, "Oh, it's nothing."

- Noticing when an idea you're reading about resonates with you and taking those extra few seconds to highlight it.

- Noticing that a note could use a better title—and changing it so it's easier for your future self to find it.

- Noticing you could move or link a note to another project or area where it will be more useful.

- Noticing opportunities to combine two or more Intermediate Packets into a new, larger work so you don't have to start it from scratch.

- Noticing a chance to merge similar content from different notes into the same note so it's not spread around too many places.

- Noticing when an IP that you already have could help someone else solve a problem, and sharing it with them, even if it's not perfect.

The nice thing about notes, unlike to-dos, is that they aren't urgent. If one important to-do gets overlooked, the results could be catastrophic. Notes, on the other hand, can easily be put on hold any time you get busy, without any negative impact. If you have the time to organize your notes each week, that's great. If you don't, it's no problem. I often will wait weeks or even a month or longer before I find the time to clear my notes inbox. They remain ready and waiting there for as long as I need.

The most common misconception about organizing I see when I'm working with clients is the belief that organizing requires a heavy lift. They seem to believe that if they could just block off their calendar and get a few days free of pressing commitments, then they'd finally be able to curb the clutter and clear their head.

Even on the rare occasions I've seen people somehow manage to clear such a big block of time, it never seems to go very well. They tend to get bogged down in minutiae and barely make a dent in the mountain of accumulated stuff they wanted to tackle. Then they're saddled with a feeling of guilt that they weren't able to make progress even with so much time at their disposal. It's not natural for humans to completely reorganize their entire world all at once. There are too many layers, too many facets of a human life, to perfectly square every little detail.

It's crucial to stay organized, but it needs to be done a little at a time in the flow of our normal lives. It needs to be done in the in-between moments of moving your projects forward as you notice small opportunities for improvement.

Here are more specific examples of what those opportunities might look like:

- You decide to visit Costa Rica on your next vacation, so you move a note with useful Spanish phrases from your "Languages" resource folder to a "Costa Rica" project folder to aid in your trip.

- Your director of engineering leaves for another job and you need to hire a new one, so you move the folder you created last time for "Engineering hire" from archives to projects to guide your search.

- You schedule the next in a series of workshops you are facilitating and move a PDF with workshop exercises from an area folder called "Workshops" to a new project folder for the specific workshop you're planning.

- You notice that you need to buy a new computer because your current one is getting too slow, so you move some articles you've saved from the "Computer research" resource folder to a new project folder called "Buy a new computer."

All these actions take mere moments, and are made in response to changes in your priorities and goals. We should avoid doing a lot of heavy lifting up front, not only because it takes up precious time and energy, but because it locks us into a course of action that might not end up being right.

When you make your digital notes a *working* environment, not just a storage environment, you end up spending a lot more time there. When you spend more time there, you'll inevitably notice many more small opportunities for change than you expect. Over time, this will gradually produce an environment far more suited to your real needs than anything you could have planned up front. Just like professional chefs keep their environment organized with small nudges and adjustments, you can use noticing habits to "organize as you go."

Your Turn: A Perfect System You Don't Use Isn't Perfect

Each of the three kinds of habits I've introduced you to—Project Kickoff and Completion Checklists, Weekly and Monthly Reviews, and Noticing Habits—are all meant to be performed quickly in the in-between spaces of your day.

They are designed to build on activities you are probably already doing in some form, adding perhaps a little bit more structure. These shouldn't be massive feats, requiring you to set aside huge chunks of time in total Zen-like isolation. That's not realistic, and if you wait until those perfect conditions arrive, you'll never take even the first step.

The checklists I've provided are a starting point to help you add some predictability in an environment that is often chaotic and unpredictable. They provide a regular cadence of actions for taking in, processing, and making use of digital information, without requiring you to stop everything and reorganize everything all at once.

I want to remind you that the maintenance of your Second Brain is very forgiving. Unlike a car engine, nothing will explode, break down, or burst into flames if you let things slide for days, weeks, or even months. The entire point of building a Second Brain and pouring your thoughts into it is to make those thoughts less vulnerable to the passage of time. They will be ready to pick up right where you left off when you have more time or motivation.

To make this concrete:

- There's no need to capture every idea; the best ones will always come back around eventually.

- There's no need to clear your inbox frequently; unlike your to-do list, there's no negative consequence if you miss a given note.

- There's no need to review or summarize notes on a strict timeline; we're not trying to memorize their contents or keep them top of mind.
- When organizing notes or files within PARA, it's a very forgiving decision of where to put something, since search is so effective as a backup option.

The truth is, any system that must be perfect to be reliable is deeply flawed. A perfect system you don't use because it's too complicated and error prone isn't a perfect system—it's a fragile system that will fall apart as soon as you turn your attention elsewhere.

We have to remember that we are not building an encyclopedia of immaculately organized knowledge. We are building a *working system*. Both in the sense that it must work, and in the sense that it is a regular part of our everyday lives. For that reason, you should prefer a system that is imperfect, but that continues to be useful in the real conditions of your life.

I. A premortem is a useful practice, similar to a postmortem used to analyze how a project went wrong, except performed *before* the project starts. By asking what is likely to go wrong, you can take action to prevent it from happening in the first place.

II. Although outside the scope of this book, I've included my recommendations for task managers on multiple operating systems in the Second Brain Resource Guide at Buildingasecondbrain.com/resources.

III. The book *Getting Things Done*, known as GTD, is a helpful counterpart to personal knowledge management, applying the same lens of "getting things off your mind" that we are using for notes to "actionable" information such as to-dos.

Chapter 10

The Path of Self-Expression

An idea wants to be shared. And, in the sharing, it becomes more complex, more interesting, and more likely to work for more people.
—adrienne maree brown, writer and activist

For most of history, humanity's challenge was how to acquire scarce information. There was hardly any good information to be found anywhere. It was locked up in difficult-to-reproduce manuscripts or stuck in the heads of scholars. Access to information was limited, but that wasn't a problem for most people. Their lives and livelihoods didn't require much information. Their main contribution was their physical labor, not their ideas.

That has all changed in just the last few decades. Historically, in the blink of an eye. Suddenly, we are all plugged into an infinite stream of data, updated continuously and delivered at light speed via a network of intelligent devices embedded in every corner of our lives.

Not only that, but the very nature of labor has changed. Value has shifted from the output of our muscles to the output of our brains. Our knowledge is now our most important asset and the ability to deploy our attention our most valuable skill. The tools of our trade have become abstract and immaterial: the building blocks of ideas, insights, facts, frameworks, and mental models.

Now our challenge isn't to acquire more information; as we saw in the exploration of divergence and convergence, it is to find ways to close off the stream so we can get something done. Any change in how we interact with information first requires a change in how we think. In this chapter we'll explore what it looks like and feels like to make that shift.

Mindset Over Toolset—The Quest for the Perfect App

The majority of this book has been about acquiring a new set of tools in your relationship to information. However, over the years I've noticed that it is never a person's toolset that constrains their potential, it's their mindset.

You might have arrived at this book because you heard about this new field called personal knowledge management, or maybe when you were trying to find guidance in how to use a cool new notetaking app. Maybe you were drawn in by the promise of new techniques for enhancing your productivity, or perhaps it was the allure of a systematic approach to creativity.

Whatever you are looking for, all these paths eventually lead to the same place, if you are willing to follow: a journey of personal growth. There is no divide between our inner selves and our digital lives: the beliefs and attitudes that shape our thinking in one context inevitably show up in other contexts as well.

Underlying our struggles and challenges with productivity, creativity, and performance is our fundamental relationship to the information in our lives. That relationship was forged during your upbringing as you encountered new experiences, and was influenced by your personality, learning style, relationships, and your genes. You learned to react in a certain way when faced with new ideas. You adopted a default "blueprint" for how you treated incoming information—with anticipation, fear, excitement, self-doubt, or some complex mix of feelings that is unique to you.

That default attitude to information colors every aspect of your life. It is the lens through which you studied for classes and took tests in school. It set the stage for the kinds of jobs and careers you pursued. At this very moment, as you read these words, that default attitude is working in the background. It is telling you what to think about what you're reading—how to interpret it, how to feel about it, and how it applies to you.

Our attitude toward information profoundly shapes how we see and understand the world and our place in it. Our success in the workforce depends on our ability to make use of information more effectively and to think better, smarter, faster. As society gets ever more complex, this emphasis on personal intelligence is only increasing. The quality of our thinking has become one of the central defining features of our identity, our reputation, and our quality of life. We are constantly advised that we need to know more to be able to achieve our goals and dreams.

What would you say if I told you that isn't true?

The Fear Our Minds Can't Do Enough

When it comes to accomplishing our goals, it's not that innate intelligence isn't valuable. I'm saying that the greater the burden you place on your biological brain to give you everything you want and need, the more it will struggle under the weight of it all. You'll feel more stressed, anxious, like there are way too many balls in the air. The more time your brain spends striving to achieve and overcome and solve problems, the less time you have left over for imagining, creating, and simply enjoying the life you're living. The brain can solve problems, but that isn't its sole purpose. Your mind was meant for much more.

It is this fundamental attitude toward information that will start to change as you integrate your Second Brain into your life. You will begin to see connections you didn't know you could make. Ideas about business, psychology, and technology will connect and spawn new revelations you've never consciously considered. Lessons from art, philosophy, and history will intermingle to give you epiphanies about how the world works. You will naturally start to combine these ideas to form new perspectives, new theories, and new strategies. You will be filled with a sense of awe at the elegance of the system you've created, and how it works in almost mysterious ways to bring to your attention the information you need.

Maybe you don't see yourself as a writer, creator, or expert. I certainly didn't when I first started taking notes on my health problems. Once you start seeing even your biggest ambitions in terms of the smaller chunks of information they are made up of, you'll begin to realize that any experience or passing insight can be valuable. Your fears, doubts, mistakes, missteps, failures, and self-criticism—it's all just information to be taken in, processed, and made sense of. All of it is part of a larger, ever-evolving whole.

A participant in one of my courses named Amelia recently told me that starting to build her Second Brain had caused her to make a 180-degree change in her relationship to the Internet. She had seen it as "sensationalistic and offensive" and, as a result, hadn't wanted to engage with the online world at all. Once she had a place where she could curate the best of the Internet while ignoring everything that didn't serve her, she told me she began to see it in a completely new light. Amelia is a skilled leadership coach who runs a clinic teaching leaders how to manage their nervous systems to improve their well-being and effectiveness. Imagine how many more people she can reach with her expertise now that she sees the Internet as a source of wisdom and connection, not just noise.

How does such a dramatic change happen? Amelia didn't necessarily learn a new fact that she didn't know before. She took on a new perspective. She chose to look at the world through a different lens—the lens of appreciation and abundance. We can't always control what happens to us, but we can choose the lens we look through. This is the basic choice we have in creating our own experience—which aspects to nourish or starve, using only the magnifying power of our attention.

As you build a Second Brain, your biological brain will inevitably change. It will start to adapt to the presence of this new technological appendage, treating it as an extension of itself. Your mind will become calmer, knowing that every idea is being tracked. It will become more focused, knowing it can put thoughts on hold and access them later. I often hear that people start to feel a tremendous sense of conviction—for their goals, their dreams, and the things they want to change or influence in the world—because they know they have a powerful system behind them amplifying every move they make.

Giving Your First Brain a New Job

Instead of trying to optimize your mind so that it can manage every tiny detail of your life, it's time to fire your biological brain from that job and give it a new one: as the CEO of your life, orchestrating and managing the process of turning information into results. We're asking your biological brain to hand over the job of remembering to an external system, and by doing so, freeing it to absorb and integrate new knowledge in more creative ways.

Your Second Brain is always on, has perfect memory, and can scale to any size. The more you outsource and delegate the jobs of capturing, organizing, and distilling to technology, the more time and energy you'll have available for the self-expression that only you can do.

Once your biology is no longer the bottleneck on your potential, you'll be free to expand the flow of information as much as you want without drowning in it. You'll be more balanced and peaceful, knowing you can step away from that flow at any time because it's all being stored safely outside your head. You will be more trusting, because you've learned to trust a system outside yourself. It will be incredibly humbling and reassuring, in fact, that you are not solely responsible for all the remembering that needs to happen in your life. You will be more open-minded, willing to consider more unorthodox, more challenging, more unfinished ideas, because you have a plentiful supply of alternatives to choose from. You'll want to expose yourself to more diverse perspectives, from more people, without necessarily committing to any single one. You'll become a curator of perspectives, free to pick and choose the beliefs and concepts that serve you best in any given situation.

Delegating a job you've been doing for a long time is always intimidating. The voice of fear creeps up in the back of your mind: "Will there be anything left for me to do?" "Will I still be valued and needed?" We are taught that it's better to have a secure role than risk being replaced. That it's safer to keep your head down and not make a fuss rather than strive for something better. Emptying ourselves of our jumble of thoughts requires courage, because without our thoughts as distractions, we are left to sit with uncomfortable questions about our future and our purpose.

That is why building a Second Brain is a journey of personal growth. As your information environment changes, the way your mind operates starts to be transformed. You leave behind one identity and step into another—an identity as the orchestrator and conductor of your life, not its passenger. Any shift in identity can feel confronting and scary. You don't know exactly who you will be and what it will be like on the other side, but if you persevere through the transition, there is always a new horizon of hope, possibility, and freedom waiting for you on the other side.

The Shift from Scarcity to Abundance

How do you know when you've begun making the shift to this new identity I've described? The biggest shift that starts to occur as soon as you start creating a Second Brain is the shift from viewing the world through the lens of scarcity to seeing it through the lens of abundance.

I see so many people trying to operate in this new world under the assumptions of the past—that information is scarce, and therefore we need to acquire and consume and hoard as much of it as possible. We've been conditioned to view information through a consumerist lens: that more is better, without limit. Through the lens of scarcity, we constantly crave more, more, more information, a response to the fear of not having enough.[1] We've been taught that information must be jealously guarded, because someone could use it against us or steal our ideas. That our value and self-worth come from what we know and can recite on command.

As we saw in the chapter on Capturing, the inclination to amass information can become an end in itself. It is all too easy to default to collecting more and more content without regard to whether it is useful or beneficial to us. This is indiscriminate consumption of information, treating every meme and random post on social media as if it was just as important as the most profound piece of wisdom. It is driven by fear—the fear of missing out on some crucial fact, idea, or story that everyone is talking about. The paradox of hoarding is that no matter how much we collect and accumulate, it's never enough. The lens of scarcity also tells us that the information we already have must not be very valuable, compelling us to keep searching externally for what's missing inside.

The opposite of a Scarcity Mindset is an Abundance Mindset. This is a way of looking at the world as full of valuable and helpful things—ideas, insights, tools, collaborations, opportunities. An Abundance Mindset tells us that there is an endless amount of incredibly powerful knowledge everywhere we look—in the content we consume, in our social network, in our bodies and intuitions, and in our own minds. It also tells us that we don't need to consume or understand all of it, or even much of it. All we need is a few seeds of wisdom, and the seeds we most need tend to continually find us again and again. You don't need to go out and hunt down insights. All you have to do is listen to what life is repeatedly trying to tell you. Life tends to surface exactly what we need to know, whether we like it or not. Like a compassionate but unyielding teacher, reality doesn't bend or cave to our will. It patiently teaches us in what ways our thinking is not accurate, and those lessons tend to show up across our lives again and again.

Making the shift to a mindset of abundance is about letting go of the things we thought we needed to survive but that no longer serve us. It means giving up low-value work that gives us a false sense of security but that doesn't call forth our highest selves. It's about letting go of low-value information that seems important, but that doesn't make us better people. It's about putting down the protective shield of fear that tells us we need to protect ourselves from the opinions of others, because that same shield is keeping us from receiving the gifts they want to give us.

The Shift from Obligation to Service

There is a second shift that occurs when you begin to use your Second Brain not only for remembering, but for connecting and creating. You will transition from doing things primarily out of obligation or pressure to doing things from a spirit of service.

I believe most people have a natural desire within them to serve others. They want to teach, to mentor, to help, to contribute. The desire to give back is a fundamental part of what makes us human.

I also notice that many people put that desire on hold. They are waiting for a future time when they will have "enough" time, bandwidth, expertise, or resources. That day seems to get continuously postponed as they get new jobs, start new careers, have kids, and simply try to keep up with the demands of life.

You are under no obligation to help others. Sometimes it's all you can do to take care of yourself. Still, I've noticed time and again a phenomenon that happens as people collect more and more knowledge in their Second Brain. That inner desire to serve slowly comes to the surface. Faced with the evidence of everything they already know, suddenly there's no longer any reason to wait.

The purpose of knowledge is to be shared. What's the point of knowing something if it doesn't positively impact anyone, not even yourself? Learning shouldn't be about hoarding stockpiles of knowledge like gold coins. Knowledge is the only resource that gets better and more valuable the more it multiplies. If I share a new way of thinking about your health, or finances, or business, or spirituality, that knowledge isn't less valuable to me. It's more valuable! Now we can speak the same language, coordinate our efforts, and share our progress in applying it. Knowledge becomes more powerful as it spreads.

There are problems in the world that you are uniquely equipped to solve. Problems in society like poverty, injustice, and crime. Problems in the economy like inequality, educational deficits, and workers' rights. Problems in organizations like retention, culture, and growth. Problems in the lives of people around you that your product or service or expertise could solve, helping them communicate, learn, or work more effectively. As Ryder Carroll says in *The Bullet Journal Method*, "Your singular perspective may patch some small hole in the vast tattered fabric of humanity."

There are people who will be reached only if they are reached by you. People who have no other source for the kind of guidance you can provide. People who don't know where to look for solutions to problems they might not even know they have. You can be that person for them. You can pay forward some of the immense care that has been poured into you by a lifetime of parents, teachers, and mentors. With mere words, you can open doors to unimaginable horizons for the people around you.

Your Second Brain starts as a system to support you and your goals, but from there it can just as easily be used to support others and their dreams. You have everything you need to give back and be a force for good in the world. It all starts with knowledge, and you have at your disposal an embarrassment of riches.

The Shift from Consuming to Creating

The practice of building a Second Brain is more than the sum of capturing facts, theories, and the opinions of others. At its core, it is about cultivating self-awareness and self-knowledge. When you encounter an idea that resonates with you, it is because that idea reflects back to you something that is already within you. Every external idea is like a mirror, surfacing within us the truths and the stories that want to be told.

In a 1966 book,[1] the British-Hungarian philosopher Michael Polanyi made an observation that has since become known as "Polanyi's Paradox." It can be summarized as "We know more than we can say."

Polanyi observed that there are many tasks we can easily perform as humans that we can't fully explain. For example, driving a car or recognizing a face. We can try to describe how we do these things, but our explanations always fall far short. That's because we are relying on *tacit knowledge*, which is impossible to describe in exact detail. We possess that knowledge, but it resides in our subconscious and muscle memory where language cannot reach.

This problem—known as "self-ignorance"—has been a major roadblock in the development of artificial intelligence and other computer systems. Because we cannot describe how we know what we know, it can't be programmed into software.

The curse of computer scientists is our blessing, because this tacit knowledge represents the final frontier on which humans outperform machines. The jobs and endeavors that rely on tacit knowledge will be the last ones to be automated.

As you build your Second Brain, you will collect many facts and figures, but they are just a means to an end: discovering the tacit knowledge that lives within you. It's in there, but you need external hooks to pull it out and into your conscious awareness. If we know more than we can say, then we need a system for continuously offloading the vast wealth of knowledge we've gained from real life experience.

You know things about how the world works that you can't fully put into words. You understand human nature at a deep intuitive level. You see patterns and connections in your field that no other machine or human can see. Life has given you a set of experiences that provide you with a unique lens on the world. Through that lens you can perceive truths that can have a profoundly positive impact on you and others.

We are constantly told that we should be true to ourselves and pursue our deepest desires, but what if you don't know what your goals and desires are? What if you have no idea what your "life purpose" is or should be? Self-direction is impossible without self-knowledge. How can you know what you want if you don't know who you are?

The process of knowing yourself can seem mystical, but I see it as eminently practical. It starts with noticing what resonates with you. Noticing what seems to call out to you in the external world and gives you a sense of déjà vu. There is a universe of thoughts and ideas and emotions within you. Over time, you can uncover new layers of yourself and new facets of your identity. You search outside yourself to search within yourself, knowing that everything you find has always been a part of you.

Our Fundamental Need for Self-Expression

In Chapter 1, I told the story of my unexplained medical condition and how it led me to start organizing information digitally.

There was a period a few years into that journey when I was at my lowest point. I had seemingly exhausted every avenue that modern medicine could offer me. The doctors were suggesting that it must be all in my head because their diagnostics couldn't find anything wrong. I was in more pain than ever before, waking up with so much tension in my neck that it felt like a vise clutching me by the throat.

I started withdrawing from my friends and social circles because I was so consumed by the pain I was experiencing. My attention was so focused on the pain in my body that I found it difficult to hold a conversation. I started spending more and more time by myself, on the Internet, where I could communicate and connect without speech. My view of life darkened as I slowly spiraled into depression and despair. It felt for a time like I had no future. How could I date or make friends without being able to speak? What kind of job could I hold down with unpredictable, chronic pain? What kind of future could I look forward to as my symptoms continued to worsen, without any treatment or even diagnosis on the horizon?

It was around this time that I made two discoveries that changed, and saved, my life. The first was meditation and mindfulness. I began to meditate and discovered a whole realm of spirituality and introspection that I never knew existed. I learned, to my astonishment, that I am not my thoughts. That my thoughts were the constant background chatter of my subconscious mind, and that I could choose whether to "believe" what they were telling me. Meditation gave me more relief from my symptoms than anything the doctors could prescribe. My pain became my teacher, showing me what needed my attention.

As I started having deep, profoundly moving experiences in meditation, I wanted to share what I was learning with others. This led me to my second great discovery: writing in public.[II] I started a blog, and my very first blog post was about my experience at a Vipassana meditation retreat in Northern California. I still had trouble speaking, so writing became my refuge. On my blog, I could share anything I wanted to, in as much detail as I wanted. I was in control, with no limits on my ability to express myself.

I discovered something through that experience: that self-expression is a fundamental human need. Self-expression is as vital to our survival as food or shelter. We must be able to share the stories of our lives—from the small moments of what happened today at school to our grandest theories of what life is about.

Your Turn: The Courage to Share

I've spoken with so many people about their stories, and I've noticed time and again how many of them have beautiful, moving, powerful things to share. They have unique experiences that have revealed to them deep wisdom, yet they almost always undervalue those stories and experiences. They think maybe one day they'll get around to sharing them. I'm here to tell you that there is no reason to wait.

The world is desperate to hear what you know. You can change lives by sharing yourself with others.

It takes courage and vulnerability to stand up and deliver your message. It takes going against the grain, refusing to be quiet and hidden in the face of fear. Finding your voice and speaking your truth is a radical act of self-worth: Who are you to speak up? Who says you have anything to offer? Who are you to demand people's attention and take up their time?

The only way to discover the answer to these questions is by speaking and seeing what comes out. Some of what you say might not resonate with others or provide value to them, but occasionally, you will strike on something—a way of seeing, a perspective, a story—that blows people's minds and visibly transforms how they see the world. It could be someone you're having coffee with, a client or customer, or your online followers. In those moments, the vast chasm that separates us as humans is bridged. For a brief moment, you get to feel in your bones that we are all in this together. We are all part of a vast tattered fabric of humanity, and your highest calling is simply to play your part in it.

With the power of a Second Brain behind you, you can do and be anything you want. Everything is just information, and you are a master at flowing and shaping it toward whatever future you desire.

Final Thoughts: You Can Do This

There is no single right way to build a Second Brain. Your system can look like chaos to others, but if it brings you progress and delight, then it's the right one.

You may start with one project and slowly move on to more ambitious or complex ones as your skills develop. Or you may find yourself using your Second Brain in completely unexpected ways that you hadn't envisioned.

As your needs change, give yourself the freedom to discard or take on whichever parts serve you. This isn't a "take it or leave it" ideology where you must accept all of it or none of it. If any part doesn't make sense or doesn't resonate with you, put it aside. Mix and match the tools and techniques you've learned in this book to suit your needs. This is how you ensure your Second Brain remains a lifelong companion through the seasons of your life.

Wherever you are at this moment—just starting a practice to consistently take notes, or finding ways to more effectively organize and resurface your best thinking, or generating more original and impactful work—you can always fall back on the four steps of CODE:

- Keep what resonates (Capture)
- Save for actionability (Organize)
- Find the essence (Distill)
- Show your work (Express)

If at any point you feel overwhelmed, take a step back and focus on what is immediately necessary: your most important projects and priorities. Scale back to only the notes you need to move those priorities forward. Instead of trying to architect your entire Second Brain system from scratch up front, focus on moving one project at a time through each step from capturing to expressing. When you do so, you'll find that the steps are much easier and more flexible than you imagined.

You can also simplify things by focusing on just one stage of building your Second Brain. Think about where you are now and where you want to be in the near future:

- Are you hoping to remember more? Focus on developing the practice of capturing and organizing your notes according to your projects, commitments, and interests using PARA.

- Are you hoping to connect ideas and develop your ability to plan, influence, and grow in your personal and professional life? Experiment with consistently distilling and refining your notes using Progressive Summarization and revisiting them during weekly reviews.

- Are you committed to producing more and better output with less frustration and stress? Focus on creating one Intermediate Packet at a time and looking for opportunities to share them in ever more bold ways.

As you begin your journey, here are twelve practical steps you can take right now to get your Second Brain started. Each one of them is a starting point to begin establishing the habits of personal knowledge management in your life:

1. **Decide what you want to capture.** Think about your Second Brain as an intimate commonplace book or journal. What do you most want to capture, learn, explore, or share? Identify two to three kinds of content that you already value to get started with.

2. **Choose your notes app.** If you don't use a digital notes app, get started with one now. See Chapter 3 and use the free guide at Buildingasecondbrain.com/resources for up-to-date comparisons and recommendations.

3. **Choose a capture tool.** I recommend starting with a read later app to begin saving any article or other piece of online content you're interested in for later consumption. Believe me, this one step will change the way you think about consuming content forever.

4. **Get set up with PARA.** Set up the four folders of PARA (Projects; Areas; Resources; Archives) and, with a focus on actionability, create a dedicated folder (or tag) for each of your currently active projects. Focus on capturing notes related to those projects from this point forward.

5. **Get inspired by identifying your twelve favorite problems.** Make a list of some of your favorite problems, save the list as a note, and revisit it any time you need ideas for what to capture. Use these open-ended questions as a filter to decide which content is worth keeping.

6. **Automatically capture your ebook highlights.** Set up a free integration to automatically send highlights from your reading apps (such as a read

later or ebook app) to your digital notes (see my recommendations at Buildingasecondbrain.com/resources).

7. **Practice Progressive Summarization.** Summarize a group of notes related to a project you're currently working on using multiple layers of highlighting to see how it affects the way you interact with those notes.

8. **Experiment with just one Intermediate Packet.** Choose a project that might be vague, sprawling, or simply hard, and pick just *one* piece of it to work on—an Intermediate Packet. Maybe it is a business proposal, a chart, a run of show for an event, or key topics for a meeting with your boss. Break the project down into smaller pieces, make a first pass at one of the pieces, and share it with at least one person to get feedback.

9. **Make progress on one deliverable.** Choose a project deliverable you're responsible for and, using the Express techniques of Archipelago of Ideas, Hemingway Bridge, and Dial Down the Scope, see if you can make decisive progress on it using only the notes in your Second Brain.

10. **Schedule a Weekly Review.** Put a weekly recurring meeting with yourself on your calendar to begin establishing the habit of conducting a Weekly Review. To start, just clear your notes inbox and decide on your priorities for the week. From there, you can add other steps as your confidence grows.

11. **Assess your notetaking proficiency.** Evaluate your current notetaking practices and areas for potential improvement using our free assessment tool at Buildingasecondbrain.com/quiz.

12. **Join the PKM community.** On Twitter, LinkedIn, Substack, Medium, or your platform(s) of choice, follow and subscribe to thought leaders and join communities who are creating content related to personal knowledge management (#PKM), #SecondBrain, #BASB, or #toolsforthought. Share your top takeaways from this book or anything else you've realized or discovered. There's nothing more effective for adopting new behaviors than surrounding yourself with people who already have them.

While building a Second Brain is a project—something you can commit to and achieve within a reasonable period of time—*using* your Second Brain is a lifelong practice. I recommend you revisit *Building a Second Brain* at various points over time. I guarantee you'll notice things you missed the first time.

Whether you focus on implementing one aspect of the CODE Method, make a full commitment to the entire process, or something in between, you are taking on a new relationship with the information in your life. You are developing a new relationship to your own attention and energy. You are committing to a new identity in which you are in charge of the information swirling around you, even if you don't always know what it means.

As you embark on the lifelong path of personal knowledge management, remember that you've achieved success before. There have been practices that you'd never heard of before, that are now integral parts of your life. There have been habits and skills that seemed impossible to master, that you now can't imagine living without. There have been new technologies that you swore you would never embrace that you now use every day. This is the same—what seems unfamiliar and strange now will eventually feel completely natural.

If I could leave you with one last bit of advice, it is to chase what excites you. When you are captivated and obsessed by a story, an idea, or a new possibility, don't just let that moment pass as if it doesn't matter. Those are the moments that are truly precious, and that no technology can produce for you. Run after your obsessions with everything you have.

Just be sure to take notes along the way.

I. *The Tacit Dimension*, by Michael Polanyi

II. I came to appreciate the incredible power of writing in public through my partnership with David Perell, who teaches people how to do it via his online writing school Write of Passage, which you can learn more about at writeofpassage.school.

Bonus Chapter

How to Create a Tagging System That Works

I wrote this book to give you a new way to think about the knowledge that matters to you. It's designed to give anyone a path to create—and benefit from—a Second Brain and provides an introduction to the fascinating world of personal knowledge management.

The practice begins and ends with notetaking—including capturing, organizing, distilling, and expressing information, ideas, and packets of work. The specific techniques in the CODE chapters are the best place to get started. However, one of the most common questions I receive is about the advanced skill of tagging.

I've compiled a bonus chapter on how to create a tagging system for your Second Brain following the principle of actionability. Although not essential for getting started, tags do provide an extra layer of organization that can be useful as your knowledge collection grows.

You can download this chapter at Buildingasecondbrain.com/bonuschapter.

Additional Resources and Guidelines

The technology landscape is constantly changing, and the best practices evolve as new platforms emerge. I've created the Second Brain Resource Guide as a public resource with continually updated recommendations for the best notes apps, capture tools and other useful apps, frequently asked questions, and other advice and guidelines to help you succeed in personal knowledge management. You can access it at https://www.buildingasecondbrain.com/resources.

Acknowledgments

I'm sitting here at my usual writing spot weeks after my manuscript deadline has passed. I've postponed the writing of these acknowledgments for as long as possible because it feels almost impossible. The number of people who have contributed to and touched this book along the way is staggering. The depth of gratitude I feel for all the love and energy and intelligence that they have poured into me is difficult to put into words. But I'll try.

Thank you to Stephanie Hitchcock and the team at Atria for your willingness to take a chance on a novel idea and a first-time author. This book exists only because you saw its potential and committed to seeing it realized. I'm deeply grateful to my editor, Janet Goldstein, for wrangling my words (and sometimes me!) into a message far clearer and more elegant than anything I could have come up with on my own. My agent, Lisa DiMona, has graciously guided me through every step of the publishing journey from the earliest days of this project. I look forward to working together for many years to come.

Thank you to the Forte Labs team and extended network—Betheny Swinehart, Will Mannon, Monica Rysavy, Marc Koenig, Steven Zen, Becca Olason, and Julia Saxena. You've been behind the scenes every step of the way making the business work, overcoming challenges, and creating new ways of sharing these ideas with the world. I am continuously amazed by your dedication to excellence. You constantly impress me with your deep commitment to creating lasting, positive change in people's lives. I look forward to everything we'll accomplish together.

I'm eternally grateful to Billy Broas for helping me find more powerful ways of communicating my truth beyond my inner circle. To Maya P. Lim for crafting the visual identity that will deliver our education to every corner of the planet. And I'm thankful to the Pen Name team for partnering with me in sharing my life's work far and wide.

The only way my business (or life) functions is with the help of my "brain trust." The work I do wouldn't be half as meaningful or interesting without your unwavering support. David, building a business with you and developing our ideas side by side has been one of the most meaningful endeavors of my career. Joel, you are like a rock in stormy seas. I've lost count of how many times dinners in your home served as a stabilizing force when it seemed like everything was going to unravel. Raphael, you came up with the name of the course—and now the book. The laughter you've brought into my life has been like a beacon of joy every time I've started to take myself too seriously. Derick, in many ways this journey started with our late-night conversations as teenagers about technology and the future—ours and humanity's. Thank you for entertaining and encouraging those far-fetched ideas, some of which, after all these years, found their way into this book.

I've had a series of mentors and advisors who shifted my trajectory in ways I can hardly believe. Thank you to Venkatesh Rao for serving as my introduction to the online world of ideas. A few words of your public support and encouragement fueled my motivation for years. Thank you to David Allen for pioneering the personal productivity field and introducing us to the possibility that we could proactively improve how we worked with and managed information. I have been profoundly influenced and helped by your ideas.

Thank you to Kathy Phelan, not only for believing I was on to something, but for sponsoring and advising me in bringing my work into companies. Your belief in me at the time outshined my own, and your advice and lessons continue to resound years later. Thank you to James Clear for giving so generously of your time and guiding my book writing efforts around many pitfalls and blind spots. At a time when everyone in the world wanted your attention, you chose to give it to a fledgling writer with not much to offer in return. Thank you to Joe Hudson, who came into my life as a friend and mentor at a crucial moment when I needed to learn how to navigate the emotions of the new level of self-expression I was taking on. Thank you to Srini Rao for going all-in with your support of my work and taking a risk in putting your reputation behind it.

Thank you to Forte Labs' followers, subscribers, customers, and students. You are the fuel powering all the capturing, organizing, distilling, and expressing that makes the *Building a Second Brain* community so vibrant. This book is just as much a distillation of the stories, strategies, and techniques I've learned from you over the years as it is my own ideas. You are the ultimate authority on what works and what doesn't. By taking my courses, reading my writing, and giving me feedback on everything from tweets to book drafts, you've opened the door to a future where Second Brains are available to people everywhere. I never expected so many people to believe in what I was doing. Every single day I have your support and attention I consider a miracle.

Everything I am ultimately comes from my family. My soil and my rock, from which the meaning and joy in my life springs. Thank you to my parents, Wayne Forte and Valeria Vassão Forte, for giving me an upbringing that exposed me to countless enriching experiences, cultures, places, and people. Dad, you are my model of what it means to express myself with unsparing honesty and taste, while also upholding my responsibilities as a father, husband, and citizen. Mom, you gave me the gifts to balance out my strong will and sharp tongue—patience, generosity, graciousness, and self-awareness. You both dedicated your lives to making me the kind of person with enough abundance to share with others. So many of the teachings expressed in this book have their origin in the simple, practical lessons you taught me and modeled for me as a child. To my siblings and in-laws, Lucas, Paloma, Marco, Kaitlyn, and Grant. You are my best friends, my confidants, and my lifelong companions. Every time I start to lose sight of who I am and what's important to me, you bring me back to the soil I came from. I cherish every minute we spend together.

And finally, from the bottom of my heart I am thankful to you, Lauren and Caio, for making all this worthwhile. Lauren, you have played every role a person can play in my life—partner, lover, cofounder, coach, advisor, and now, wife and mother. You became whoever you needed to become, acquired whichever skills were demanded of you, and ventured into one new territory after another, all to help me reach my dreams. There's nothing more gratifying in my life than watching you grow and evolve into the most inspiring, genuine, open-hearted person I've ever met. I consider it my highest privilege to walk alongside you as you step into your greatness. Caio, you've only just arrived, but already I can't live without you. You make my life so much more colorful and hilarious. My love for you gives me the determination to become the best version of myself I can. My highest hope for this book is that it makes the world a safer, more humane, and more interesting place for you.

CHAPTER 1: WHERE IT ALL STARTED

1 Erik Brynjolfsson and Andrew McAfee, *The Second Machine Age: Work, Progress, and Prosperity in a Time of Brilliant Technologies* (New York: W. W. Norton & Company, 2014), Amazon Kindle Location 1990 of 5689.

CHAPTER 2: WHAT IS A SECOND BRAIN?

1. Nick Bilton, "Part of the Daily American Diet, 34 Gigabytes of Data," *New York Times*, December 9, 2009, https://www.nytimes.com/2009/12/10/technology/10data.html.

2. Daniel J. Levitin, "Hit the Reset Button in Your Brain," *New York Times*, August 9, 2014, https://www.nytimes.com/2014/08/10/opinion/sunday/hit-the-reset-button-in-your-brain.html?smprod=nytcore-iphone&smid=nytcore-iphone-share.

3. Microsoft, *The Innovator's Guide to Modern Note Taking: How businesses can harness the digital revolution*, https://info.microsoft.com/rs/157-GQE-382/images/EN-US%2017034_MSFT_WWSurfaceModernNoteTaking_ebookRefresh_R2.pdf.

4. IDC Corporate USA, *The Knowledge Quotient: Unlocking the Hidden Value of Information Using Search and Content Analytics*, http://pages.coveo.com/rs/coveo/images/IDC-Coveo-white-paper-248821.pdf.

5. Robert Darnton, *The Case for Books: Past, Present, and Future* (New York: PublicAffairs, 2009), 224.

6. Craig Mod, "Post-Artifact Books and Publishing," craigmod.com, June 2011, https://craigmod.com/journal/post_artifact/.

7. Including innovators like Paul Otlet, Vannevar Bush, Doug Engelbart, Ted Nelson, and Alan Kay, among many others.

CHAPTER 3: HOW A SECOND BRAIN WORKS

1 Wikipedia, s.v., "Molecular Structure of Nucleic Acids: A Structure for Deoxyribose Nucleic Acid," accessed October 13, 2021, https://en.wikipedia.org/wiki/Molecular_Structure_of_Nucleic_Acids:_A_Structure_for_Deoxyribose

2 Deborah Chambers and Daniel Reisberg, "Can mental images be ambiguous?," *Journal of Experimental Psychology: Human Perception and Performance* 11, no. 3 (1985): 317–28, https://doi.org/10.1037/0096-1523.11.3.317.

3 Nancy C. Andreasen, "Secrets of the Creative Brain," July/August 2014, https://www.theatlantic.com/magazine/archive/2014/07/secrets-of-the-creative-brain/372299/.

4 Wikipedia, s.v., "Recency Bias," accessed October 13, 2021, https://en.wikipedia.org/wiki/Recency_bias.

5 Robert J. Shiller, "What to Learn in College to Stay One Step Ahead of Computers," *New York Times*, May 22, 2015, https://www.nytimes.com/2015/05/24/upshot/what-to-learn-in-college-to-stay-one-step-ahead-of-computers.html?smprod=nytcore-iphone&smid=nytcore-iphone-share.

6 For a fascinating look into how persuasion and sales is becoming a fundamental part of almost everyone's job, see Daniel Pink, *To Sell Is Human: The Surprising Truth About Moving Others* (New York: Penguin Group, 2012), 6.

7 Tim Ferriss, *Tools of Titans: The Tactics, Routines, and Habits of Billionaires, Icons, and World-Class Performers* (New York: HarperCollins, 2017), 421.

8 Each of these stories is real, but names have been changed to protect their anonymity.

9 Erwin Raphael McManus, *The Artisan Soul: Crafting Your Life into a Work of Art* (New York: HarperCollins, 2014), 171.

CHAPTER 4: CAPTURE—KEEP WHAT RESONATES

1 Wikipedia, s.v., "Taylor Swift," accessed October 13, 2021, https://en.wikipedia.org/wiki/Taylor_Swift.

2 Swiftstyles II, "Taylor Swift being a songwriting genius for 13 minutes," July 27, 2020, YouTube video, 13:52, https://www.youtube.com/watch?v=bLHQatwwyWA.

3 NME, "Taylor Swift—How I Wrote My Massive Hit 'Blank Space,'" NME.com, October 9, 2015, YouTube video, 3:58, https://www.youtube.com/watch?v=8bYUDY4lmls.

4 Gian-Carlo Rota, *Indiscrete Thoughts* (Boston: Birkhäuser Boston, 1997), 202.

5 James Gleick, *Genius: The Life and Science of Richard Feynman* (New York: Open Road Media, 2011), 226.

6 Raymond S. Nickerson, "Confirmation Bias: A Ubiquitous Phenomenon in Many Guises," *Review of General Psychology 2*, no. 2 (June 1998): 175–220, https://journals.sagepub.com/doi/10.1037/1089-2680.2.2.175.

7 Marianne Freiberger, "Information is surprise," *Plus Magazine*, March 24, 2015, https://plus.maths.org/content/information-surprise.

8 Dacher Keltner and Paul Ekman, "The Science of 'Inside Out,'" *New York Times*, July 3, 2015, https://www.nytimes.com/2015/07/05/opinion/sunday/the-science-of-inside-out.html.

9 Stephen Wendel, *Designing for Behavior Change: Applying Psychology and Behavioral Economics* (Sebastopol, CA: O'Reilly Media, 2013).

10 Zachary A. Rosner et al., "The Generation Effect: Activating Broad Neural Circuits During Memory Encoding," *Cortex* 49, no. 7 (July–August 2013), 1901–1909, https://doi.org/10.1016/j.cortex.2012.09.009.

11 James W. Pennebaker, "Writing about Emotional Experiences as a Therapeutic Process," *Psychological Science* 8, no. 3 (May 1997), 162–66.

CHAPTER 5: ORGANIZE—SAVE FOR ACTIONABILITY

1 Twyla Tharp, *The Creative Habit: Learn It and Use It For Life* (New York: Simon & Schuster, 2003), 80.

2 Joan Meyers-Levy and Rui Zhu, "The Influence of Ceiling Height: The Effect of Priming on the Type of Processing That People Use," *Journal of Consumer Research* 34, no. 2 (2007): 174–86, https://doi.org/10.1086/519146.

3 Adam Davidson, "What Hollywood Can Teach Us About the Future of Work," *New York Times Magazine,* May 5, 2015.

CHAPTER 6: DISTILL—FIND THE ESSENCE

1. "Inside Francis Ford Coppola's *Godfather* Notebook": https://www.hollywoodreporter.com/news/general-news/inside-francis-ford-coppolas-godfather-notebook-never-before-seen-photos-handwritten-notes-9473-947312/.

2. AFI's 100 Years...100 Movies - 10th Anniversary Edition, Television Academy.

3. *Francis Coppola's Notebook*, imdb.com, 2001, https://www.imdb.com/title/tt0881915/.

4. Jess Wise, "How the Brain Stops Time," *Psychology Today*, March 13, 2010, https://www.psychologytoday.com/us/blog/extreme-fear/201003/how-the-brain-stops-time.

5. Meghan Telpner, *Academy of Culinary Nutrition*, Academy of Culinary Nutrition (blog), https://www.culinarynutrition.com/blog/.

6. Artyfactory. (n.d.). Animals in Art—Pablo Picasso. Retrieved January 27, 2022, from https://www.artyfactory.com/art_appreciation/animals_in_art/pablo_picasso.htm.

CHAPTER 7: EXPRESS—SHOW YOUR WORK

1 Octavia E. Butler, *Bloodchild and Other Stories: Positive Obsession* (New York: Seven Stories, 2005), 123–36.

2 Lynell George, *A Handful of Earth, A Handful of Sky: The World of Octavia Butler* (Santa Monica: Angel City Press, 2020).

3 Dan Sheehan, "Octavia Butler has finally made the *New York Times* Best Seller list," LitHub.com, September 3, 2020, https://lithub.com/octavia-butler-has-finally-made-the-new-york-times-best-seller-list/.

4 Butler's archive has been available to researchers and scholars at the Huntington Library since 2010.

5 Deborah Barreau and Bonnie A. Nardi, "Finding and Reminding: File Organization from the Desktop," *ACM SIGCHI Bulletin* 27, no. 3 (1995), 39–43, https://doi.org/10.1145/221296.221307. Joseph A. Maxwell, "Book Review: Bergman, M. M. (Ed.). (2008). Advances in Mixed Method Research. Thousand Oaks, Ca: Sage," *Journal of Mixed Methods Research* 3, no. 4 (2009), 411–13, https://doi.org/10.1177/1558689809339316.

6 William P. Jones and Susan T. Dumais, "The spatial metaphor for user interfaces: experimental tests of reference by location versus name," *ACM Digital Library* 4, no. 1 (1986), https://doi.org/10.1145/5401.5405.

7 Adam Savage, "Inside Adam Savage's Cave: Model Making for Movies," Adam Savage's Tested, YouTube video, 20:26, https://www.youtube.com/watch?v=vKRG6amACEE.

CHAPTER 8: THE ART OF CREATIVE EXECUTION

1 Danny Choo, "DIY: How to write a book," boingboing, January 27, 2009, https://boingboing.net/2009/01/27/diy-how-to-write-a-b.html.

CHAPTER 9: THE ESSENTIAL HABITS OF DIGITAL ORGANIZERS

1 Dan Charnas, *Work Clean: The Life-Changing Power of Mise-en-Place to Organize Your Life, Work, and Mind* (Emmaus, PA: Rodale Books, 2016).

CHAPTER 10: THE PATH OF SELF-EXPRESSION
1 Lynne Twist, *The Soul of Money* (New York City: W. W. Norton & Company, 2017), 43.

Index

A

Abhinavagupta, 33
abstract thinking, 160n
actionability, 45–46, 101–3
    to avoid getting stuck, 86–90
    kitchen organization analogy, 103–5
    priming for, 90–95
Afrofuturism, 148
Ahrens, Sönke, 141n
Allen, David, 9, 212
Amazon Kindle, 75
American Film Institute, 114
Andreasen, Nancy C., 35
Apple Notes, 39
Archipelago of Ideas, 182–86, 192, 205, 240
Archives (PARA category), 87, 89, 107–8
    defined, 90
    explained, 95
    illustration of folder, 99
    kitchen organization analogy, 103–4, 105
    moving projects to across all platforms, 207, 209–10
    ordering by actionability, 102, 103
Areas (PARA category), 87, 89
    defined, 90
    examples of, 93
    explained, 92–94
    illustration of folder, 98
    kitchen organization analogy, 104, 105
    ordering by actionability, 102, 103
    reviewing, 216
Arieti, Silvano, 175
Atlantic, The, 15

Atomic Habits (Clear), 197
attention
    decreased quality of, 26-27
    investments in, 202
    poverty of, 18, 18n
    protecting, 149–51
audio/voice transcription apps, 73, 74, 130–32

B

basic notes apps, 73, 74
batch processing of notes, 212. *See also* Monthly Reviews; Weekly Reviews
"bicycle for the mind" metaphor, 3
biji, 20n
biological brain
    increased cognitive demands on, 4, 18
    a new job for, 227–29
"Blank Space" (song), 55
bookmarks, 59
box, the, 81–84, 114, 178
brain. *See* biological brain; Second Brain
brown, adrienne maree, 223
browsing, 160–61
Buildingasecondbrain.com/bonuschapter, 162n, 243
Buildingasecondbrain.com/course, 16n
Buildingasecondbrain.com/quiz, 240
Buildingasecondbrain.com/resources, 40, 66, 74n, 208n, 239, 240
Building a Second Brain system. *See also* Second Brain
    as a journey of personal growth, 228–29
    legacy of thinkers and creators behind, 16
    lessons of, 3
    major influence on, 2–3
    origins of, 15–16
    overview, 2–5
Bullet Journal Method, The (Carroll), 232
bullet points, 125
Burns, Ken, 136–37, 137n
Bush, Vannevar, 4n, 30n
Butler, Octavia Estelle, 19, 145–49, 178

C

calendar, checking, 213
campsite rule, 139–40
Capture (CODE step), 42, 43, 53–79, 116–17, 238

    building a private knowledge collection with, 54–56
    choosing tools for, 73–76
    creating a knowledge bank with, 57–60
    criteria for, 66–70
    current thinking on a project, 203–4
    deciding on content, 239
    divergence and, 180
    explained, 44–45
    inspiring material, 68
    Organize step separated from, 101–2, 117
    personal information, 69
    resonance and, 44–45, 70–72
    surprising benefits of, 76–78
    surprising information, 69–70
    "Twelve Favorite Problems" exercise, 61–66, 239
    useful information, 68–69
    what if it were easy?, 78–79
    what not to keep, 60–61
Carroll, Ryder, 232
case files, 57
Case for Books, The (Darnton), 19–20
Cathedral Effect, 84–86
Challenger space shuttle disaster, 62
Chambers, Deborah, 35
China, 20n
choreography, 81–84
citing sources, 154, 169
Clear, James, 197
code libraries, 57
CODE Method, 33, 42–49, 180, 198, 237–38. *See also* Capture; Distill; Express; Organize
cognitive ability enhancement, 3–4, 5
commonplace books, 43, 57, 147, 178
    Archipelago of Ideas for, 183
    digital, 21–22
    legacy of, 19–21
computer desktop, clearing, 213–14
confirmation bias, 69
connect (Express stage), 165–66
connecting (PKM stage), 41, 42
consumption-to-creation shift, 48, 232–34
content, 57–58
convergence, 181
    example of, 191–94

features of, 178–80
Coppola, Francis Ford, 113–16, 178, 179
Covey, Stephen, 186n
COVID-19 pandemic, 147
create (Express stage), 166–67
creating (PKM stage), 41, 42
creative execution, 175–96
    Archipelago of Ideas in, 182–86, 192, 205, 240
    convergence and divergence in (*see* convergence; divergence)
    Dial Down the Scope technique in, 188–91, 194, 196, 240
    Hemingway Bridge in, 186–88, 193, 196, 210, 240
Creative Habit, The (Tharp), 81
creative process, 177
creativity, 48–49
    attention in, 149
    Capture step and, 54–56, 57, 65, 72
    cause of blocks in, 38
    collaborative nature of, 167–68
    connecting ideas in, 35
    from consumption to, 48, 232–34
    incubation of ideas in, 36–37
    productivity and, 197
    as a remix, 169
Creativity: The Magic Synthesis (Arieti), 175
Crick, Francis, 34–35
Curator's Perspective, 67, 204, 228

D

dance choreography, 81–84
Darnton, Robert, 19–20
deliverables, 153, 154, 240
Designing for Behavior Change (Wendel), 71–72
Design Thinking, 178n
detachment gain, 77n
Detachment Gain, The: The Advantage of Thinking Out Loud (Reisberg), 77n
Dewey decimal system, 45, 86
Dial Down the Scope technique, 188–91, 194, 196, 240
digital garden, 4n
digital hoarding, 54
discoverability, 118–19
Distill (CODE step), 42, 43, 113–43, 238
    convergence and, 180
    discoverability and, 118–19

    explained, 46–47
    Progressive Summarization in (*see* Progressive Summarization)
    pruning the good to surface the great, 134–37
    quantum notetaking and, 116–18
    for your future self, 117, 118–19, 122, 125, 139, 141–43
distilled notes, 153
divergence, 181
    example of, 191–94
    features of, 178–80
DNA structure, discovery of, 34–35
documents created by others, 153, 154
Dropbox, 110

E

ebook apps, 73, 75
80 percent done rule, 166n
Einstein, Albert, 47
email
    capturing excerpts from, 76
    clearing the inbox, 213
essence, finding, 46–47
Evernote, 39, 40n
executive summary, 124–26
Express (CODE step), 42, 43, 145–72, 238
    convergence and, 180
    explained, 47–49
    Intermediate Packets in (*see* Intermediate Packets)
    protecting attention, 149–51
    as a remix, 169–70
    synonyms for, 49, 49n
    three stages of, 164–67
    you only know what you make, 170–72
extended mind, 30, 30n, 31
Extended Mind, The (Paul), 17, 30n
external knowledge, 58–59

F

Fast Company, 15
favorites, 59
fear
    of mind limitations, 225–27
    of missing out (FOMO), 195
feedback, 168

Feynman, Richard, 61–63
film making, 113–16, 169
final deliverables, 153, 154
Flaubert, Gustave, 81
folders. *See also* PARA
    browsing, 160–61
    defined, 88n
    moving relevant notes to, 203, 205
    reviewing for relevant notes, 203, 204
    searching for related terms across, 203, 204–5
FOMO (Fear of Missing Out), 195
Forte, Wayne Lacson, 175–76
four-letter frameworks, 87n
Francis Coppola's Notebook (documentary), 115
Franklin, Rosalind, 34
Frequently Asked Questions (FAQs) pages, 161–62
future self, 2
    Capture step and, 67
    Distill step and, 117, 118–19, 122, 125, 139, 141–43
    Organize step and, 88

G

Genentech, 15
Generation Effect, 76–77
Genius: The Life and Science of Richard Feynman (Gleick), 62–63
Getting Things Done (Allen), 9, 212, 212n
Gleick, James, 62–63
goals
    moving to "Completed" section, 207, 208–9
    PARA and, 89
    reviewing and updating, 215
    stress-free accomplishment of, 166–67
Godfather, The (film), 113–16, 179
Google Docs, 40n, 110
Google Keep, 39
grid code, 43n

H

habits, 197–221
    mise-en-place process and, 198–200
    noticing, 200, 217–20
    project checklists, 200, 201–11
    reviews (*see* Monthly Reviews; Weekly Reviews)

  using an imperfect system, 220–21
Harvard Business Review, 15
Harvard University Library, 19
heavy lift approach, 36
Hemingway, Ernest, 186
Hemingway Bridge, 186–88, 193, 196, 210, 240
Hidalgo, César, 58n
highlighting, 75, 118–19. *See also* Progressive Summarization
  automatically capturing ebook, 239–40
  excessive, 137–38
  making it difficult, 140–41
  without a purpose in mind, 138–40
highlights (insightful passages), 58
Hofstra College, 114
Hollywood model, 91
Hollywood Reporter, 114
hook books, 57
"How the Brain Stops Time" *(Psychology Today)*, 121, 123, 124, 125
How to Take Smart Notes (Ahrens), 141n
human capital, 14
Huntington Library, 148

I

ideas
  connecting together, 35, 41
  designing a space for, 84–86
  incubating over time, 36–37
  making concrete, 34–35
  retrieving exactly when needed, 164–65
  revealing new associations between, 35–36
IDEO, 178n
images, 59, 163
Inc., 15
Industrial Light & Magic, 169
Industrial Revolution, 19
inflection point, 30
information
  average person's daily consumption of, 17
  deluge of, 17–18
  personal, 69
  sensitive, 60–61
  surprising, 69–70
  time spent searching for by average employee, 18

    transformation to knowledge, 48
    useful, 68–69
information hoarding, 2, 229–30
insights, 59
inspiration, 68
Inter-American Development Bank, 15
Intermediate Packets (IPs), 151–64, 166–68, 170, 187, 194, 195, 196, 238
    assembling the building blocks of, 155–58
    defined, 152
    experimenting with just one, 240
    five types of, 153
    as intellectual property, 152n
    retrieving, 158–64
    reviewing and moving to other folders, 207, 209
internal knowledge, 59
International Data Corporation, 18
intuition, 71–72
IPs. *See* Intermediate Packets
Iriki, Atsushi, 22n
Is This Anything? (Seinfeld), 56

J
Japan, 20n
Jobs, Steve, 3n
Joel, Billy, 82–83
Johnson, Steven, 182
Junger, Sebastian, 38

K
kitbashing, 169
kitchen organization analogy, 103–5
knowledge. *See also* personal knowledge management
    building a private collection of, 54–56
    external, 58–59
    information transformed to, 48
    internal, 59
    notes as building blocks of, 22–25
    radically expanding the definition of, 57–58
    recycling, 201–2
    tacit, 233
knowledge assets, 58
knowledge bank, 57–60
knowledge garden, 54, 59, 159

Knowledge Management, 2n. *See also* personal knowledge management
knowledge map, 126–27
knowledge vault, 21
knowledge workers
    attention and, 149, 150
    defined, 22–23
    designing an environment for, 85
    Dial Down the Scope technique and, 189

L

Lao Tzu, 113
Le Cunff, Anne-Laure, 4n
LEGO analogy, 24, 65, 152
Leonardo da Vinci, 19
Locke, John, 19, 122n
Lucas, George, 114
luck, engineering, 217–20
Luhmann, Niklas, 4n

M

Magic Number (4), 87n
Maravita, Angelo, 22n
marginalia, 21, 21n
meditation, 11, 235
meeting notes, 59, 132–34
Memex, 4n, 30n
memory(ies), 18–19, 59. *See also* remember; remembering
Microsoft, 18
Microsoft OneNote, 39
Microsoft Word, 40n
mindfulness, 235
mise-en-place process, 198–200
Mod, Craig, 21
Monthly Reviews, 200, 212, 215–17
Moser, Edvard, 43n
Moser, May-Britt, 43n
musings, 59
MythBusters (television program), 169

N

New Method of Making Common-Place Books, A (Locke), 122n
New York Times, 17, 91
Nobel Prize winners, 61, 62, 63, 186
Norwegian University of Science and Technology, 43n

notes
    batch processing of, 212
    creating an outline of, 203, 205
    discoverability of, 118–19
    distilled, 153
    as knowledge building blocks, 22–25
    meeting, 59, 132–34
    moving to project folders, 203, 205
    paper, 38n
    reviewing folders for relevant, 203, 204
    telling a bigger story with, 165–66
    where to save individual, 101–3
notetaking
    assessing proficiency of, 240
    author's experience, 10–14
    quantum, 116–18
    rethinking, 22–25
    three most common novice mistakes, 137–41
notetaking apps
    browsing features, 160–61
    choosing, 38–40, 239
    clearing notes in, 214
    content not suited for, 60–61
    exporting and importing features, 40n
    four powerful characteristics of, 39
    search function in, 159–60
noticing habits, 200, 217–20
Notion, 39

O

obligation-to-service shift, 231–32
official kickoff, 206–7
online articles, 129–30
open-ended questions, 64, 66, 239
Organize (CODE step), 42, 43, 81–111, 238
    actionability and (*see* actionability)
    Capture step separated from, 101–2, 117
    Cathedral Effect and, 84–86
    divergence and, 180
    explained, 45–46
    move quickly, touch lightly approach, 108–11
    PARA in (*see* PARA)
    where to save individual notes, 101–3

outlines, 184–85, 203, 205
outtakes, 153

P

PARA, 87–111, 160, 163, 165, 204, 214, 238. *See also* Archives; Areas; Projects; Resources
    creative execution and, 194
    getting set up with, 239
    how it works, 90–95
    illustrations of, 96–101
    kitchen organization analogy, 103–5
    versatility of, 87, 89
    Weekly Reviews and, 212
Parable of the Sower, The (Butler), 147–48
Paramount Pictures, 113
Paul, Annie Murphy, 17, 30n
Pauling, Linus, 34
Peace Corps, 12–13
Perell, David, 235n
perfectionism, 40, 89
personal information, 69
personal knowledge management (PKM), 2–3, 2n, 4, 224
    joining the community, 240–41
    three stages of, 41–42
perspectives, sharpening, 37–38
physical products, 34–35, 58n
Picasso, Pablo, 134–36
Picasso's Bull (drawing), 134–36, 137
pillow books, 20n
PKM. *See* personal knowledge management
podcast player apps, 75
podcasts, 130–32
Polanyi, Michael "Polanyi's Paradox," 233
postmortem questions, 210
poverty of attention, 18, 18n
premortem questions, 206, 206n
Princeton University, 37
productivity, 48, 48n
    creativity and, 197
    mise-en-place process and, 198–200
Progressive Summarization, 120–34, 137–43, 153, 163, 184, 204, 205, 238
    defined, 120
    examples of, 128–34
    four layers of, 120–26, 120n

    knowledge map navigation with, 126–27
    practicing, 240
    three common novice mistakes, 137–41
project checklists, 200, 201–11
Project Completion Checklist, 207–11
Project Kickoff Checklist, 202–7
project lists, reviewing and updating, 216
Projects (PARA category), 87, 89
    adding a current status note to inactive, 208, 210
    completed as oxygen of Second Brain, 105–8
    defined, 90
    80 percent done rule, 166n
    examples of, 91
    explained, 90–92
    illustrations of folders, 96–97, 100–101
    kitchen organization analogy, 104, 105
    marking as complete, 207, 208
    officially closing and celebrating, 211
    ordering by actionability, 102, 103
    stress-free completion of, 166–67
prompt books, 114–16, 178
Psychology Today article, 121, 123, 124, 125

Q

Quantified Self community, 11n
quantum notetaking, 116–18
questions
    open-ended, 64, 66, 239
    postmortem, 210
    premortem, 206, 206n
    "Twelve Favorite Problems" exercise, 61–66, 239
quotes, 58

R

reactivity loop, 77–78
read later apps, 73, 74, 75, 239
recency bias, 36–37
reflections, 59
Reisberg, Daniel, 35, 77n
remember (Express stage), 164–65
remembering (PKM stage), 41, 42
remodeling project example, 97, 132–34, 191–94
resonance, 44–45, 70–72

Resources (PARA category), 87, 89
    defined, 90
    explained, 94–95
    illustration of folder, 98–99
    kitchen organization analogy, 104, 105
    ordering by actionability, 102, 103
retrieval
    exactly when needed, 164–65
    methods, 158–64

S

Savage, Adam, 169
scarcity-to-abundance shift, 229–30
search
    for ease of retrieval, 159–60
    of outlines, 185
    for related terms across all folders, 203, 204–5
Second Brain. *See also* Building a Second Brain system
    biological brain changed by, 227
    comparison of life with and without, 25–29
    completed projects as oxygen of, 105–8
    defined, 4
    engineering luck with, 217–20
    information stream filtered by, 53–54
    shifts in mindset and, 229–34
    superpowers of, 34–38
    synonyms for, 4, 4n
    twelve steps to starting, 238–41
Seinfeld, Jerry, 56
selection, 185
self-expression, 223–41
    the courage to share and, 236–37
    fundamental need for, 234–36
    mindset over toolset in, 224–25
    a new job for the biological brain, 227–29
    shifts in mindset and, 229–34
self-ignorance, 233
sensitive information, 60–61
sequencing, 185
serendipity, 162–64
Shannon, Claude, 69
sharing, 231–32
    courage for, 236–37

    discovering the power of, 14–15
    serendipity and, 163–64
"She's Got a Way" (song), 83
Simon, Herbert, 18n
slow burn approach, 37
social media apps, 73, 74
songwriting, 54–56, 57
sources, citing, 154, 169
space, designing for ideas. *See* Cathedral Effect
spatial thinking, 160n
stakeholders, communicating with, 206, 211
standardization, 177
Stanford Design School, 178n
stigmergy, 139n
stories, 59
success criteria
    defining, 206
    evaluating, 211
Swift, Taylor, 54, 178
swipe files, 57

T

Tacit Dimension, The (Polanyi), 233n
tacit knowledge, 233
tagging, 88, 88n
    creating a system that works, 243
    for ease of retrieval, 161–62
    reviewing tags for relevant notes, 203, 204
takeaways, 59
task manager, 207, 208
tasks
    choosing for the week, 214–15
    reprioritizing, 217
    reviewing someday/maybe, 216–17
Telpner, Meghan, 131–32
Tharp, Twyla, 81–84, 114, 178
thinking
    abstract, 160n
    benefits of externalizing thoughts, 76–78
    emotions' effect on, 71
    spatial, 160n
    technology as a tool for, 3, 30–31, 41
    technology-induced changes in, 19

Toyota, 15
Tversky, Barbara, 160n
"Twelve Favorite Problems" exercise, 61–66, 239

V

Vico, Giambattista, 145, 170
voice memos, 59, 75

W

Watson, James, 34–35
Wayback Machine, 72n
Wayne Lacson Forte: On My Way To Me (documentary), 176n
web clipper apps, 73, 74
Weekly Reviews, 102, 200, 212–15, 240
Why Information Grows (Hidalgo), 58n
Wikipedia, 118n, 128–29
Wilkins, Maurice, 34
Woolf, Virginia, 19
work-in-process, 153, 154
Write of Passage (online school), 235n
writing in public, 235–36
writing things down, 2. *See also* notetaking
    discovering the power of, 10–14
    multiple benefits of, 76–78

Y

YouTube videos, capturing parts of, 76

Z

Zettelkasten, 4n
zuihitsu, 20n

www.ingramcontent.com/pod-product-compliance
Lightning Source LLC
Chambersburg PA
CBHW081616100526
44590CB00021B/3456